Green Energy and Technology

For further volumes:
http://www.springer.com/series/8059

José Ramón San Cristóbal Mateo

Multi-Criteria Analysis in the Renewable Energy Industry

 Springer

José Ramón San Cristóbal Mateo
Escuela Técnica Superior de Náutica
University of Cantabria
Santander
Spain
e-mail: jose.sancristobal@unican.es

ISSN 1865-3529   e-ISSN 1865-3537
ISBN 978-1-4471-6081-6   ISBN 978-1-4471-2346-0 (eBook)
DOI 10.1007/978-1-4471-2346-0
Springer London Dordrecht Heidelberg New York

British Library Cataloguing in Publication Data
A catalogue record for this book is available from the British Library

Printed on acid-free paper

Springer is part of Springer Science+Business Media (www.springer.com)

Preface

Sustainable Development is defined as meeting the needs of the current generation without compromising the ability of future generations to meet their own needs. Nowadays, Sustainable Development has been the subject of wide-ranging discussions and debates within government, non-government and academic circles, being a major focus of national and international economic, social and environmental agendas so that a good quality of life can be enjoyed by current and future generations. However, Sustainable Development is threatened by industrial pollution emissions which cause environmental problems such as deforestation, soil erosion, acid rain, toxicity, water pollution, depletion of fisheries and the expansion of desert areas. The problem of climate change, which occurs due to the increasing concentrations of greenhouse gasses in the atmosphere, has caused great concern at all levels, from the general public to national governments and international agencies. Renewable energy sources can be an important remedy to many environmental problems that the world faces today. Some new governmental policies have been adopted to encourage the introduction of energy efficiency measures, the technical changes and the renewable and sustainable energy. The energy planning scenario has completely changed over the las two decades, from an almost exclusively concern with cost minimization of supply-side options to the need of explicitly multiple and potentially conflicting objectives. In addition, different and numerous groups of actors, such as institutions and administration authorities, potential investors and environmental groups, get involved in the process. This complex environment indicates the multi-criteria character of the problem. The traditional single criteria decision-making approach is no longer able to handle these problems. The policy formulation for fossil fuels energy substitution by renewable energies must be addressed in a multi-criteria context. Multi-criteria decision-making is a decision support approach that is suitable for addressing complex problems featuring high uncertatinty, conflicting objectives, different forms of data and information and multi interests and perspectives. The multiplicity of criteria and the complexity of energy planning and energy projects makes multi-criteria analysis a valuable tool in the decision-making process. It is the aim of this book to show the use of the multi-criteria decision-making methods to provide valuable assistance in reaching equitable and acceptable solutions in the selection of renewable energy projects. This book is aimed at those actors who bring along different criteria and point of view, have interests at stake and therefore, influence the decision-making process.

Cantabria, Spain, November 2011   José Ramón San Cristóbal Mateo

Contents

Chapter 1
The Renewable Energy Industry and the Need for a Multi-Criteria Analysis

Energy systems play an important role in the economic and social development of a country. The selection among different alternatives for the expansion of renewable energy sources, where the environmental dimension is added to the usual economic, technical and social criteria demands the utilization of powerful decision-aid techniques. The traditional single decision-making approach is no longer able to handle these problems. The multiplicity of criteria, and the involvement of different actors in the decision procedure, makes multi-criteria analysis a valuable tool in the policy formulation for fossil fuel energy substitution by renewable energy sources.

1.1 Introduction

Renewable energy (RE) is the energy generated from natural resources such as sunlight, wind, rain, biomass, tides and geothermal heat. The exploitation of RE sources has gained a vast interest during the last years. Global environmental problems, including the greenhouse effect and ozone holes, which have a negative impact on the health and living quality of humans, are now identified as significant problems. It is now widely recognized that the largest source of atmospheric pollution stems from fossil fuel combustion, upon which current energy production and use patterns throughout the world rely heavily [1]. The energy demand of fossil fuel has major consequences on the environment. The emission of toxic chemical pollutants, greenhouse gasses, and other pollutants cause climate change and environmental pollution of air, land and water. In addition, the growing rise in the cost of raw materials driven up by oil prices, which are strongly affected by political conditions in the world, can have a negative effect on the world economies. In this context, the raising awareness of environmental issues, the precarious nature of dependency on fossil fuel imports, the vulnerability of many states as regard energy sector and the advent of RE alternatives, have forced many countries, specially the developed ones, to use RE sources.

J. R. San Cristóbal Mateo, *Multi-Criteria Analysis in the Renewable Energy Industry*,
Green Energy and Technology, DOI: 10.1007/978-1-4471-2346-0_1,
© Springer-Verlag London Limited 2012

Clean, domestic and renewable energy is commonly accepted as the key for future life [2]. Setting goals to ensure that the environment is safeguarded and the effects of climate change are mitigated while seeking long-lasting Sustainable Development is one of the most important issues faced by the world today [3]. The Sustainable Development has been the subject of wide-ranging discussion and debate within government, non-government and academic cycles, being a major focus of national and international economic, social and environmental agendas. Sustainable Development is defined as meeting the needs of the current generation without compromising the ability of future generations to meet their own needs [4]. Energy systems play an important role in the economic and social development of a country and the living of people. A balance of energy production and consumption which has no, or minimal, negative impact on the environment, but gives the opportunity for a country to employ its social and economic activities can be seen as the final goal [5].

The use of renewable energy sources for the production of electric power brings huge benefits, both in terms of environmental protection as well as savings in non-renewable resources [3]. Renewable energy sources are environment-friendly compared to fossil fuels, with low levels or absence of both air pollutants and greenhouse gasses emissions, waste valuation potential and capable of replacing conventional sources in a variety of applications at competitive prices [6]. A more intensive use of renewable energy sources by means of improvements in energy performance and transport, conversion and conservation technologies, as well as development of new energy sources, can be an important remedy to many environmental problems that the world faces today. Some new governmental policies have been adopted to encourage the introduction of energy efficiency measures, the technical changes and the renewable and sustainable energy. The Kyoto Protocol, the EU Renewable Directive 2001/77/EC [7] and the European Biomass Action Plan [8] are examples of ambitious political goals fostering the development of conversion technologies based on Renewable energy sources.

1.2 The Need for a Multi-Criteria Analysis

Given that the energy sector and energy planning in particular affects the interests and resources of multiple actors, it is socially not acceptable to suggest (or even implement) a policy alternative without taking into account the interests and preferences of the (multiple) affected factors [9]. Different and numerous groups of actors get involved in the process, each group brings along different criteria and point of view, which must be resolved within a framework of understanding and mutual compromise (concessions) [10]. The actors include those groups of individuals, institutions and administration authorities such as local authorities and communities, potential investors, academic institutions, environmental groups, governments, that through their priorities and evaluation systems, have interests at stake and directly or indirectly influence the decision-making process.

Table 1.1 Main criteria in the renewable energy industry [5]

Criteria	Description
Technical criteria	
Efficiency	It is the ratio of the output energy to the input energy
Exergy efficiency	Exergy is the energy that is available to be used
Primary energy ratio (PER)	PER is defined as the ratio of consumption primary energy to the users' demand energy
Safety	Safety can be considered an evaluation criteria to show the effects of energy systems to society and people
Reliability	It may be defined as the capacity of a device or system to perform as designed
Maturity	It measures the degree of maturity of the technology showing how widespread the technology is at both national and international level
Environmental criteria	
Particles emissions	It involves mainly NO_x, CO_2, CO and SO_2 emissions which can lead to global warming and contribute to the greenhouse effect
Non-methane volatile organic compounds (NMVOCs)	NMVOC is a generic term for a large variety of chemically different compounds, like for example, benzene, ethanol, formaldehyde, cyclohexane or acetone
Land use	The environment and landscape are affected directly by the land occupied by energy systems. Quality of people's life is affected by energy systems as it could have been used for creation of parks and recreation centers. Particularly energy supply systems with biomass and biofuels require a large amount of land
Noise	Noise pollution form energy systems is displeasing machine-created sound that disrupts the activity or balance of human and animal life
Topography	It may play an important role when selecting between several alternatives such as wind farm locations

Another factor that has completely changed the energy planning scenario over the last two decades, from an almost exclusively concern with cost minimization of supply-side options to the need of explicitly multiple and potentially conflicting objectives is the increasingly complex social, economical, technological, and environmental scenario. Multiple criteria that affect the success of a renewable energy project must be analyzed and taking into account. The literature traditionally divides the used criteria into four aspects technical, environmental, economical, and social as shown in Tables 1.1 and 1.2 [5].

The traditional single criteria decision making approach, normally aimed at identifying the most efficient option at a low cost, is no longer able to handle these problems. From an interdisciplinary point of view, exploiting renewable energy sources (potential solutions, courses of actions, decision alternatives, selecting the best

Table 1.2 Main criteria in the renewable energy industry [5]

Criteria	Description
Economic criteria	
Investment cost	Purchase of mechanical equipment, technological installations, construction of roads and connections to the national grid, engineering services, drilling and other incidental construction work
Operation and maintenance costs	Operation cost such as employees' wages, the funds spent for the energy, the products and services for the energy system operation and maintenance costs that aim to prolong energy systems life and avoid failures
Fuel costs	It refers to the fund spent for the provision of raw material necessary for energy supply system operation
Electric cost	It is the product cost of power plant, to evaluate its economic performance from the viewpoint of consumers
Internal rate of return	It is a commonly used investment criteria expressing the rate of return of a project
Net present value	It is defined as the total present value of a time series of cash flows
Payback period	Payback period refers to the period of time required for the return on an investment to 'repay' the sum of the original investment
Service life	It is the expected lifetime of an energy system
Equivalent annual cost	Equivalent annual cost (EAC). It is the cost per year of owning and operating an asset over its entire life span. It is calculated by dividing the NPV of an energy project by the present value of an annuity factor
Social criteria	
Social acceptability	It is a qualitative criteria expressing the overview of opinions related to the energy systems by the local population regarding the hypothesized realization of the projects under review from the consumer point of view
Job creation	Number of jobs created during the project's cycle life
Social benefits	It expresses the social progress in the local regions that house the energy systems, estimating the total social and economic impact that may become perceptible by introducing an energy project

alternative, identifying non-sustainable energy supply system, informing design-makers of the integrated performances of the alternatives and monitoring impacts on the social environment, concentrate investments in the most appropriate areas) can no longer be based on a single-dimensional axis of evaluation, such as cost or benefit. The policy formulation for fossil fuels energy substitution by RE must be addressed in a multi-criteria context. The multiplicity of criteria and the complexity of energy planning and energy projects make multi-criteria analysis a valuable tool in the decision-making process. Decision-makers need to examine more than one objective in reaching the final goal set with respect to the satisfaction of constraints set by the environment.

References

1. Madlener R, Antunes CH, Dias LC (2009) Assessing the performance of biogas plants with multi-criteria and data envelopment analysis. Eur J Oper Res 197:1084–1094
2. Kahraman C, Kaya I, Cebi S (2009) A comparative analysis for multiattribute selection among renewable energy alternatives using fuzzy axiomatic design and fuzzy analytic hierarchy process. Energy 34:1603–1616
3. Cavallaro F (2010) A comparative assessment of thin-film photovoltaic production processes using the ELECTREIII method. Energ Policy 38:463–474
4. World Commission on Environment and Development (1987) Our common future. Oxford University Press, Oxford
5. Wang JJ, Jing YY, Zhang CF et al (2010) Review on multi-criteria decision analysis aid in sustainable energy decision-making. Renew Sust Energ Rev 13:2263–2278
6. Aras H, Erdogmus S, Koc E (2004) Multi-criteria selection for a wind observation station location using analytic hierarchy process. Renew Energ 23:1383–1392
7. Directive 2001/77/EC (27.09.2001), Official J Eur Commun L 283/33 (27.10.2001)
8. National Biomass Action Plan (2005). European Comission COM(2005)628
9. Tsoutos T, Drandaki M, Frantzeeskaki N et al (2009) Sustainable energy planning by using multi-criteria analysis application in the island of Crete. Energ Policy 37:1587–1600
10. Haralambopoulos A, Polatidis H (2003) Renewable energy projects: structuring a multi-criteria group decision-making framework. Renew Energ 28:961–973

addressed in a multi-criteria context. The multiplicity of criteria and the complexity of energy planning and energy projects make multi-criteria analysis a valuable tool in the decision-making process. Decision-makers need to examine more than one objective in bearing the final goal ... with respect to the satisfaction of constraints set by the stockholders.

References

[Reference list too faded/reversed to transcribe reliably]

Chapter 2
Multi-Criteria Analysis

Decision-making is the study of identifying and choosing alternatives to find the best solution based on different factors and considering the decision-makers' expectations. Every decision is made within a decision environment, which is defined as the collection of information, alternatives, values and preferences available at the time when the decision must be made. The difficult point in decision-making is the multiplicity of the criteria set for judging the alternatives. The objectives are usually conflicting and, in most of the cases, different groups of decision-makers are involved in the process. To facilitate this type of analysis, a family of tools referred to as Multi-criteria decision-making methods gained ground due to the need to have a formalized method to assist decision-making in situations involving multiple criteria.

2.1 Introduction

Multi-criteria decision-making methods is a branch of a general class of Operations Research models that is suitable for addressing complex problems featuring high uncertainty, conflicting objectives, different forms of data and information, multi interests and perspectives, and the accounting for complex and evolving biophysical and socio-economic systems [1]. This major class of methods is further divided into Multi-objective decision-making and Multi-attribute decision-making. These methodologies share the common characteristics of conflict among criteria, incommensurable units and difficulties in design/selection of alternatives. The main distinction between the two groups of methods is based on the number of alternatives under evaluation. Multi-attribute decision-making methods are designed for selecting discrete alternatives while Multi-objective decision-making methods are more adequate to deal with multi-objective planning problems, when a theoretically infinite number of continuous alternatives are defined by a set of constraints on a vector of decision variables [2–4]. In Multi-objective decision-making (also known as multi objective programming or a vector optimization/maximization/minimization problem), the alternatives are not predetermined but instead a set of objective

J. R. San Cristóbal Mateo, *Multi-Criteria Analysis in the Renewable Energy Industry*,
Green Energy and Technology, DOI: 10.1007/978-1-4471-2346-0_2,
© Springer-Verlag London Limited 2012

functions are optimized subject to a set of constraints. The most satisfactory and efficient solution is sought. In this identified efficient solution it is not possible to improve the performance of any objective without degrading the performance of at least one other objective. In Multi-attribute decision-making a small number of alternatives are to be evaluated against a set of attributes which are often hard to quantify [5].

Following Tsoutos et al. [6] there are four starting reasons that justify the use of Multi-criteria decision-making methods: (i) It allows for investigation and integration of the interests and objectives of multiple actors since the input of both quantitative and qualitative information from every actor is taken into account in form of criteria and weight factors; (ii) It deals with the complexity of the multi-actor setting by providing output information that is easy to communicate to actors. The user-friendliness of the method lies on two aspects: the suggested criteria are estimated and given values that are consistent and comparable with the input data (as a measure of appropriateness); and the 'simple' format of the output of the method that makes the method's results meaningful and directly applicable for the interested actors; (iii) It is well-known and applied method of alternatives' assessment that also includes different versions of the method developed and researched for specific problems and/or specific contexts and (iv) It is a method that allows for objectivity and inclusiveness of different perceptions and interests of actor without being energy and cost intensive.

These methods can provide solutions to increasing complex management problems. They provide better understanding of inherent features of decision problem, promote the role of participants in decision-making processes, facilitate compromise and collective decisions and provide a good platform to understand the perception of models and analysts in a realistic scenario. The methods help to improve quality of decisions by making them more explicit, rational and efficient. Negotiating, quantifying and communicating the priorities are also facilitated with the use of these methods [7].

It should be noted that methods and results are not necessarily comparable. Every method has its restrictions, mostly due to model assumptions, which should be considered when the method is used. Inconsistences might arise because [8]: (i) the choice problem formulations do not reflect the same preference structures, (ii) the ways in which preference information is processed vary between different methods and (iii) the methods interpret the criterion weights differently.

2.2 Model Building

Multi-criteria decision-making may be considered as a complex and dynamic process including one managerial level and one engineering level. The managerial level defines the goals, and chooses the final optimal alternative whereas the the engineering level defines the alternatives, points out the consequences of choosing any one of them from the stand point of various criteria and performs the multi-criteria ranking of alternatives. The engineering level performs the optimization procedure. At the managerial level, public officials, called decision-makers have the power to accept or reject the solution proposed by the engineering level [9]. The decision-making process usually includes five main stages: defining the problem, generating alternatives and establishing criteria, criteria selection, criteria weighting, evaluation, selecting the appropriate multi-criteria method and finally ranking the alternatives. The main steps of Multi-criteria decision-making are the following:

Step 1. Defining the problem, generating alternatives and establishing criteria

A decision-making problem should start out by clearly defining the problem, discerning the alternatives, identifying the actors, the objectives and any points in conflict, together with the constraints, the degree of uncertainty and the key issues. After this, the problem can be framed indicating the evaluation criteria.

Step 2. Assigning criteria weights

The next steps include the assignment of criteria weights. These weights, that show the relative importance of criteria in the multi-criteria problem under consideration, can be determined by techniques such as Analytical Hierarchy Process and Sismos approach.

Step 3. Construction of the evaluation matrix

The phase in which the model is built constitutes a process from which the 'essence' of the problem is extracted from the complex picture drawn up so that the problem can be assessed adequately. At the end of this step, the MCDM problem can be expressed in matrix form as follows:

$$
\begin{array}{ll}
\text{Criteria} & C_1, C_2, \ldots, C_n \\
\text{Weights} & W_1, W_2, \ldots, W_n \\
\text{Alternatives} &
\end{array}
$$

$$
\begin{bmatrix} A_1 \\ A_2 \\ \vdots \\ A_m \end{bmatrix}
\begin{bmatrix}
x_{11} & x_{12} & \cdots & x_{1n} \\
x_{21} & x_{22} & \cdots & x_{2n} \\
\vdots & \vdots & \vdots & \vdots \\
x_{m1} & x_{m2} & \cdots & x_{mn}
\end{bmatrix} \tag{2.1}
$$

where x_{ij} is the evaluation given to alternative ith with respect to criterion jth, w_j is the weight of criteria j, n is the number of criteria and m is the number of alternatives.

Step 4. Selecting the appropriate method

A multi-criteria method must be selected and applied to the problem under consideration in order to rank alternatives. The data and the degree of uncertainty are key factors for the decision-maker when selecting among several multi-criteria methods.

Step 5. Ranking the alternatives

Finally, the alternatives' ranking is ordered and the best ranked alternative is proposed as a solution.

References

1. Wang JJ, Jing YY, Zhang CF et al (2010) Review on multi-criteria decision analysis aid in sustainable energy decision-making. Renew Sust Energ Rev 13:2263–2278
2. Korhonen P, Moskowitz H, Wallenius J (1992) Multiple criteria decision support-A review. Eur J Oper Res 63(3):361–375
3. Hayashi K (2000) Multicriteria analysis for agricultural resource management: a critical survey and future perspectives. Eur J Oper Res 122:486–500
4. Belton V, Stewart T (2002) Muliple criteria decision analysis: an integrated approach. Kluwer Academic, Dordrecht
5. Hwang CL, Yoon K (1981) Multiple attribute decision making: methods and applications. Springer, Berlin
6. Tsoutos T, Drandaki M, Frantzeeskaki N et al (2009) Sustainable energy planning by using multi-criteria analysis application in the island of crete. Energ Policy 37:1587–1600
7. Pohekar SD, Ramachandran M (2004) Application of multi-criteria decison-making to sustainable energy planning-A review. Renew Sust Energ Rev 8:365–381
8. Kangas J, Kangas A (2002) Multiple criteria decision support methods in forest management. An overview and comparative analyses. In: Pukkala T (ed.). Multi-objective forest planning. Kluwer Academic Publishers. Managing Forest Ecosystems 6:37–70
9. Opricovic S, Tzeng GH (2004). Compromise solution by MCDM methods: a comparative analysis of VIKOR and TOPSIS. Eur J Oper Res 156:445–455

Chapter 3
AHP

When, for a given problem, a decision-maker wishes to evaluate the performance of a number of alternative solutions and multiple objectives are important, it may be difficult to choose between alternatives. These alternatives can be evaluated in terms of a number of decision criteria and sometimes an alternative is superior in terms of one or some of the decision criteria, but inferior in terms of some other criteria. There are several reasons for adopting the AHP method as a multi criteria decision-making tool. For instance, qualitative criteria, such as political or social factors, need to be included in the decision-making process; a large quantity of criteria need to be considered or, indeed, when dealing with future technologies, characteristics of these technologies may not be well known.

3.1 Introduction

The Analytical Hierarchy Process (AHP) is a decision-making procedure which was originally developed by Saaty in the 1970s [1], and since then has been an effective tool in structuring and modeling situations involving multiple and sometimes conflicting objectives. The objective of using AHP is to identify the preferred alternative and also determine a ranking of the alternatives when all the decision criteria are considered simultaneously.

AHP is applied by breaking down an unstructured problem into components parts. Hierarchical orders are arranged by forming hierarchical structures with goal (objective) at the top of the hierarchy, criteria affecting the decision at the intermediate level, and decisions options (alternatives) at the bottom of the hierarchy. The decision-maker is then guided through a series of pair-wise comparison judgments to express relative strength or intensity of impact of the elements in the hierarchy. These judgments, represented by assigning numerical values, are then synthesized in the use of eigen vectors to determine which variables have the highest priority.

J. R. San Cristóbal Mateo, *Multi-Criteria Analysis in the Renewable Energy Industry*, Green Energy and Technology, DOI: 10.1007/978-1-4471-2346-0_3, © Springer-Verlag London Limited 2012

Table 3.1 Integer values and interpretation

a_{ij} Value	Interpretation
1	Objectives i and j are of equal importance
3	Objective i is weakly more important than objective j
5	Experience and judgements indicate that objective i is strongly more important than objective j
7	Objective i is very strongly or demonstrably more important than objective j
9	Objective i is absolutely more important than objective j
2, 4, 6, 8	Intermediate values, for example, a value of 8 means that objective i is midway between strongly and absolutely more important than objective j

AHP is an intuitive method for formulating and analyzing decisions which uses the following four steps in solving a problem: Structuring the decision problem into a hierarchical model, Obtaining the weights for each criteria, Finding the score of each alternative for each criteria and Obtaining and overall score for each alternative.

Step 1. Structuring of the decision problem into a hierarchical model

Each energy technology has many components and its use has social, environmental, political or economical implications. These are sometimes dependent on each other. The first step involves structuring of the decision into a hierarchical model built with support from specialists in applicable fields. This involves the decomposition of the problem into elements according to their common characteristics and forming a hierarchical model at different levels showing the relationships between the goal, criteria (subcriteria if necessary) and alternatives.

Step 2. Obtaining the weights for each criteria

In the second step, once the relevant criteria and alternatives have been determined together with the relations between them, the elements of a particular level are compared pair-wise with respect to a specific element in the immediate upper level. A judgmental matrix, the pairwise comparison matrix (A), is formed using linguistic terms that include the verbal judgments of the decision-maker. The entry in row i and column j of A (called a_{ij}) indicates how much more important criterion i is than j with respect to the alternative. To transform the verbal judgments into numerical quantities, Saaty [1] suggested the use of integer values constituting a 1–9 scale as shown in Table 3.1.

When constructing the pairwise comparison matrix, the following rules must be verified:

1. If $a_{ij} = \alpha$, then $a_{ji} = \frac{1}{\alpha}$.
2. If criteria i is judged to be of equal relative importance as criteria j, then $a_{ij} = a_{ji} = 1$, and $a_{ii} = 1$ for all i.

Table 3.2 RI for different values of n

n	2	3	4	5	6	7	8	9	10	11	12
RI	0	0.58	0.90	1.12	1.24	1.32	1.41	1.45	1.49	1.51	1.48

3. If all the comparisons are perfectly consistent, then the relation $a_{ik} = a_{ij} = a_{jk} \forall i, j, k$.

To recover the vector $W = [w_1, w_2, \ldots, w_N]$ from A, which indicates the weight that each criteria is given in the pairwise comparison matrix, the following two-step procedure is used:

1. For each of the As column divide each entry in column i of A by the sum of the entries in column i. This yields a new matrix, called A_{norm} (for normalized) in which the sum of the entries in each column is 1.
2. Estimate W_i as the average of the entries in row i of A_{norm}.

Once we have the pairwise comparisons matrix it is necessary checking it for consistency. Slight inconsistencies are common and do not cause serious difficulties. We can use the following four-step procedure to check for the consistency on the decision-maker's comparisons.

1. Compute AW^T, where W denotes our estimate of the decision-maker's weight.
2. Find out the maximum Eigen value (λ_{max}):

$$\frac{1}{n}\sum_{i=1}^{n} \frac{i\text{th entry in } AW^T}{i\text{th entry in } W^T} \tag{3.1}$$

3. Compute the Consistency Index (CI) as follows:

$$CI = \frac{(\lambda_{max}) - n}{n - 1} \tag{3.2}$$

The smaller the CI, the smaller the deviation form the consistency is. If CI is sufficiently small, the decision-maker's comparisons are probably consistent enough to give useful estimates of the weights for their objective. For a perfectly consistent decision-maker, the ith entry in $AW^T = n(i\text{th entry of } W^T)$. This implies that a perfectly consistent decision-maker has CI = 0.

4. Compare the Consistency Index to the Random Index (RI) for the appropriate value of n, shown in Table 3.2.

If $\frac{CI}{RI} < 0.10$, the degree of consistency is satisfactory, but if $\frac{CI}{RI} > 0.10$, serious inconsistencies may exist, and the AHP may not yield meaningful results.

Step 3. Finding the score of each alternative for each criteria

We now determine how well each alternative 'satisfies' or 'scores' for each objective. To determine these scores, a pairwise comparison matrix for each objective must be

	Criteria
Table 3.3 Criteria to evaluate a wind farm location	
c_1	Topography
c_2	Operation and maintenance cost
c_3	Land use
c_4	Infrastructure
c_5	Investment cost

constructed. In this matrix, the rows and columns are the decision-maker's possible decisions.

Step 4. Obtaining overall score for each alternative

Synthesis is the process of weighting and combining priorities through the model that leads to the overall results. Through this process, we synthesize the objective weights with the scores of each alternative on each criteria to obtain an overall score of each alternative. For calculating the 'composite weight', the following equation is used:

$$\text{Final priority of alternative } 1 = \sum_i \frac{\text{(Local priority of } A1 \text{ with respect to } C1)}{\times \text{(Local priority of } Ci \text{ with respect to goal)}}$$

$$(3.3)$$

3.2 Application

In this section, the AHP is applied to the selection of the most convenient location for a wind farm project in the autonomous region of Cantabria, in the north of Spain. Four locations, *La Braguía* (A_1), *Cilda* (A_2), *San Pedro* (A_3) and *Estacas* (A_4) will be evaluated according to the criteria shown in Table 3.3.

A typical AHP structure based on a three-level hierarchy is shown in Fig. 3.1 where the top most level represents the main goal or focus of the problem. The intermediate levels correspond to the criteria and sub-criteria, while the lowest level contains the decision alternatives.

In the selection of the RE alternatives, the following pairwise comparison matrix indicates how much more important criterion i (say, Topography) is than criterion j (say, Operation and Maintenance cost). Thus, $a_{1,2} = 5$ indicates that Topography is strongly more important than Operation and Maintenance Cost.

$$A = \begin{bmatrix} 1 & 5 & 9 & 3 & 5 \\ \frac{1}{5} & 1 & 5 & \frac{1}{3} & \frac{1}{3} \\ \frac{1}{9} & \frac{1}{5} & 1 & \frac{1}{5} & \frac{1}{7} \\ \frac{1}{3} & 3 & 5 & 1 & 1 \\ \frac{1}{5} & 3 & 7 & 1 & 1 \end{bmatrix}$$

Fig. 3.1 Hierarchical AHP model

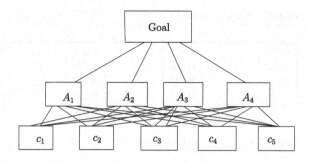

Following the two-step procedure (Step 2), we obtain the weights for each criteria: Each entry in column i of A is divided by the sum of the entries in column i, which yields the matrix A_{norm}. Next, W is estimated as the average of the entries in row i of A_{norm}.

$$W = \begin{bmatrix} w_1 & w_2 & w_3 & w_4 & w_5 \end{bmatrix} = \begin{bmatrix} 0.50 & 0.10 & 0.03 & 0.19 & 0.19 \end{bmatrix}$$

Applying Eq. 3.1 in order to check for consistency, the Eigen value obtained is $\lambda_{max} = 5.29$ and the Consistence Ratio is 0.065, which is less than the allowed value of 0.1. Thus, there is a good consistency in the judgments made.

With respect to Topography, Operation and Maintenance cost, Land use, Infrastructure and Investment cost, the following comparison matrices show how well each alternative 'scores' for each objective:

Topography	Oper. and maint. costs	Land use	Infrastructure	Investment cost
$\begin{bmatrix} 1 & 6 & 3 & 4 \\ \frac{1}{6} & 1 & \frac{1}{6} & \frac{1}{2} \\ \frac{1}{3} & 6 & 1 & 3 \\ \frac{1}{4} & 2 & \frac{1}{3} & 1 \end{bmatrix}$	$\begin{bmatrix} 1 & 2 & 4 & 6 \\ \frac{1}{2} & 1 & 2 & 3 \\ \frac{1}{4} & \frac{1}{2} & 1 & 2 \\ \frac{1}{6} & \frac{1}{3} & \frac{1}{2} & 1 \end{bmatrix}$	$\begin{bmatrix} 1 & \frac{1}{3} & \frac{1}{5} & \frac{1}{6} \\ 3 & 1 & \frac{1}{6} & \frac{1}{2} \\ 5 & 6 & 1 & 3 \\ 6 & 2 & \frac{1}{3} & 1 \end{bmatrix}$	$\begin{bmatrix} 1 & \frac{1}{3} & \frac{1}{6} & \frac{1}{5} \\ 3 & 1 & \frac{1}{3} & \frac{1}{2} \\ 6 & 3 & 1 & \frac{1}{2} \\ 5 & 2 & 2 & 1 \end{bmatrix}$	$\begin{bmatrix} 1 & \frac{1}{3} & \frac{1}{5} & \frac{1}{2} \\ 3 & 1 & \frac{1}{3} & 3 \\ 5 & 3 & 1 & 4 \\ 2 & \frac{1}{3} & \frac{1}{4} & 1 \end{bmatrix}$

Table 3.4 shows the weight vector for each alternative obtained from the comparison matrices and Table 3.5 shows the Eigen value (λ_{max}) and the Consistence Ratio for each criterion. All the Consistence Ratios are less than the allowed value of 0.1. Thus, there is a good consistency in the judgments made.

Following step 4, we now synthesize the (criteria) weights with the score of each alternative on each criteria to obtain an overall score for each alternative (Table 3.6). Applying Eq. 3.3 *La Braguía* is the best ranked alternative, followed by *Cilda, San Pedro*, with the *Estacas* alternative being the least preferred alternative.

Table 3.4 Weight vectors

c_1	c_2	c_3	c_4	c_5
$\begin{bmatrix} 0.527 \\ 0.064 \\ 0.291 \\ 0.117 \end{bmatrix}$	$\begin{bmatrix} 0.519 \\ 0.260 \\ 0.140 \\ 0.081 \end{bmatrix}$	$\begin{bmatrix} 0.064 \\ 0.128 \\ 0.552 \\ 0.256 \end{bmatrix}$	$\begin{bmatrix} 0.064 \\ 0.170 \\ 0.347 \\ 0.419 \end{bmatrix}$	$\begin{bmatrix} 0.083 \\ 0.257 \\ 0.532 \\ 0.128 \end{bmatrix}$

Table 3.5 Eigen value and consistence ratio for each criteria

	c_1	c_2	c_3	c_4	c_5
λ_{max}	4.12	4.01	4.21	4.14	4.11
$\frac{CI}{RI}$	0.046	0.004	0.079	0.051	0.040

Table 3.6 Overall score for each alternative

Alternative	Score
A_1 La Braguía	0.343
A_2 Cilda	0.341
A_3 San Pedro	0.176
A_4 Estacas	0.141

3.3 Concluding Remarks

AHP is a decision-making tool that works well when the available data for the decision-makers are difficult to quantify or when it is necessary to rely on expert opinions in some intangible aspects. However, the method has been frequently criticized for its inability to adequately accommodate the inherent uncertainty and imprecision associated with certain environments. In addition, it is difficult to subjectively scale a concrete quantitatively number to enable the pairwise comparison without losing some degree of accuracy. Despite these drawbacks, the relative ease with which it handles multiple criteria and performs qualitative and quantitative data has favored its use as a decision-making method.

AHP has been widely used in various domains related to the renewable energies. Ramanathan and Ganesh [2] performed a multiple objectives analysis to evaluate energy alternatives for lighting in households by employing goal programing and AHP. Mohsen and Akash [3] evaluated solar water heating systems in comparison with other conventional systems in Jordan and Akash et al. [4] performed a comparison between the different electricity power production options in Jordan. Wang and Feng [5] discussed the relationship between China's rural energy and its sustainable development calculating the weighting of an index system for appraising the sustainable development of rural energy using AHP. Nigim et al. [6] used AHP to assist communities in the pre-feasibility ranking of alternative RE projects to integrate the use of RE sources with conventional fuel sources in order to minimize the dependence on imported energy sources and the negative environmental impact of generating energy. Aras et al. [7] determined the most convenient location for a wind farm observation station to be built on the campus of a university. Kablam [8] presented an AHP-based modeling framework for the prioritization of energy conservation policy instruments in Jordan and Vashishtha and Ramachandran [9] evaluated demand side management strategies for reducing capacity shortages and improving system load factors in the Indian power sector from a multiobjective perspective using AHP. Nagesha and Balachandra [10] identified and prioritized relevant barriers to energy efficiency in small-scale industry clusters based on the perceptions and experience of entrepreneurs and Kon Lee et al. [11] determined the priorities in technology development for the energy efficiency and greenhouse gas control plans in Korea. Quintero et al. [12] performed a comparative analysis of the technological options using different feedstocks for fuel ethanol production and Kon Lee et al. [13] evaluated the potential of Korea to be competitive in development of hydrogen energy technology. Lee et al. [14] proposed a MCDM model based on AHP associated with benefits, opportunities, costs and risks to help select a suitable wind farm project in China.

References

1. Saaty T (1980) The analytical hierarchy process. McGraw-Hill, New York
2. Ramanathan R, Ganesh LS (1995) Energy alternatives for lighting in households: an evaluation using an integrated goal programming-AHP model. Energy 20(1):63–72
3. Mohsen MS, Akash BA (1997) Evaluation of domestic solar water heating system in Jordan using analytical hierarchy process. Energy Convers Manage 38(18):1815–1822
4. Akash BA, Mamlock R, Mohsen MS (1999) Multi-criteria selection of electric power plants using analytical hierarchy process. Electr Power Syst Res 52:29–35
5. Wang X, Feng Z (2002) Sustainable development of rural energy and its appraisal system in China. Renewable Sustainable Energy Rev 6:395–404
6. Nigim K, Munier N, Green J (2004) Pre-feasibility MCDM tools to aid communities in prioritizing local viable renewable energy sources. Renew Energy 29:1775–1791
7. Aras H, Erdogmus S, Koc E (2004) Multi-criteria selection for a wind observation station location using analytic hierarchy process. Renew Energy 23:1383–1392
8. Kablam MM (2004) Decision support for energy conservation promotion: an analytic hierarchy process approach. Energy Policy 32(10):1151–1158
9. Vashishtha S, Ramachandran M (2006) Multi criteria evaluation of demand side management (DSM) implementation strategies in the Indian power sector. Energy 31:2210–2225
10. Nagesha N, Balachandra P (2006) Barriers to energy efficiency in small industry clusters: multicriteria-based prioritization using the analytic hierarch process. Energy 31(12): 1633–1647
11. Kon Lee S, Yoon YJ, Kim JW (2007) A study on making a long-term improvement in the national energy efficiency and GHG control plans by the AHP approach. Energy Policy 35(5):2862–2868
12. Quintero JA, Montoya MI, Sánchez OJ, Giraldo OH, Cardona CA (2008) Fuel ethanol production form sugarcane and corn: comparative analysis for a Colombian case. Energy 33:385–399
13. Kon Lee S, Mogi G, Kim JW (2008) The competitiveness of Korea as a developer of hydrogen energy technology: the AHP approach. Energy Policy 36(4):1284–1291
14. Lee AHI, Chen HH, Kang HY (2009) Multi-criteria decision making on strategic selection of wind farms. Renew Energy 34:120–126

Chapter 4
Weighted Sum Method and Weighted Product Method

In this chapter we look at two simple multi-criteria decision-making methods, the Weighted Sum method and the Weighted Product method. In the Weighted Sum method the score of an alternative is equal to the weighted sum of its evaluation ratings, where the weights are the importance weights associated with each attribute. In the Weighted Product method, instead of calculating sub-scores by multiplying performance scores times attribute importance, performance scores are raised to the power of the attribute importance weight.

4.1 Weighted Sum Method

The Weighted Sum method is a straightforward method, especially used in single dimensional problems. If there are m alternatives and n criteria, then the best alternative is the one that satisfies the following expression:

$$A^*_{wsm} = Max \sum_i^j a_{ij} w_j \tag{4.1}$$

for $i = 1, 2, \ldots, m$ where A^*_{wsm} is the weighted sum method score of the best alternative, n is the number of decision criteria, a_{ij} is the actual value of the ith alternative in terms of the jth criterion and w_j is the weight of importance of the jth criterion. The total value of each alternative is equal to the sum of products. Difficult with this method merges when it is applied to multi-dimensional decision-making problems. In combining different dimensions, and consequently different units, the additive utility assumption is violated [1].

4.2 Weighted Product Method

The Weighted Product method is similar to the Weighted Sum method. The main difference is that instead of addition in this model there is multiplication. Each alternative is compared with the others by multiplying a number of ratios, one for

J. R. San Cristóbal Mateo, *Multi-Criteria Analysis in the Renewable Energy Industry,*
Green Energy and Technology, DOI: 10.1007/978-1-4471-2346-0_4,
© Springer-Verlag London Limited 2012

Table 4.1 Evaluation matrix

Alternative	Power ($Gwh/year$)	NPV($Euros * 10^6$)	Maturity(1–5)	Emissions(tCO_2/y)	Jobs
A_1 Energetic cultivations	1,709	7.50	2	2.25	15
A_2 Forest and agricultural wastes	1,815	8.65	5	2.65	22
A_3 Farming industrial wastes	698	9.30	4	2.46	18
A_4 Forest industrial wastes	698	9.65	3	3.10	25

each criterion. Each ratio is raised to the power equivalent to the relative weight of the corresponding criterion. In general, in order to compare the alternatives A_k and A_l the following product is obtained:

$$R\left(\frac{A_k}{A_l}\right) = \prod_{j=1}^{n} \left(\frac{a_{kj}}{a_{lj}}\right)^{w_j} \tag{4.2}$$

where n is the number of criteria, a_{ij} is the actual value of the ith alternative in terms of the jth criterion and w_j is the weight of importance of the jth criterion. If $R(\frac{A_k}{A_l})$ is greater than one, then alternative A_k is more desirable than alternative A_l (in the maximization case). The best alternative is the one that is better than or at least equal to the other alternatives [2].

4.3 Application

In this section, to show the use of the Weighted Sum and Weighted Product methods, the selection among four different Biomass plants for electric generation shown in Table 4.1. is considered. The alternatives will be evaluated according to the following criteria: Electric generation ($Gwh/year$), Net Present Value (NPV), Maturity (M), tons of emissions of CO_2 avoided per year (tCO_2/y) and number of Jobs created (J).

By applying Eq. (4.1) the weighted sum method scores of the four alternatives are calculated. From Table 4.2. alternative A_4, a Biomass plant with forest and industrial wastes is considered the best ranked alternative by the Weighted sum method.

In order to compare the alternatives by the Weighted Product Method, Eq. (4.2) is applied. The results, shown in Table 4.3, indicate that alternative A_2, a Biomass plant with forest and agricultural wastes with $R = 1.60$, is now the best alternative.

Table 4.2 Weighted sum method results	Alternative			
	A_1			2,475,173
	A_2			2,860,186
	A_3			3,036,073
	A_4			3,205,074

Table 4.3 Weighted product method results	Alternative	A_1	A_2	A_3	A_4
	A_1	1	0.63	1.20	1.15
	A_2	1.60	1	1.21	1.27
	A_3	1.32	0.83	1	1.05
	A_4	1.26	0.79	0.95	1

4.4 Concluding Remarks

Applications of the Weighted Sum method to sustainable energy systems can be found in Jovanovic et al [3], Afgan and Carvalho [4, 5, 6], Liposcak et al. [7], Pilavachi et al [8], Begic and Afgan [9], Afgan et al. [10], Patlitzianas et al [11] and Renn [12]. Due to the simplicity of the Weighted Sum Method, it often lies behind numerous other much more complex aggregation methods such as AHP. With regard to the Weighted Product Method a major drawback is that it over-values extremes leading to undesirable results because it considerably favors/disfavors the final evaluation of any alternative which, relative to one criterion, is far from the average.

References

1. Solnes J (2003) Environmental quality indexing of large industrial development alternatives using AHP. Environ Impact Asses 23(3):283–303
2. Chang YH, Yeh CH (2001) Evaluating airline competitiveness using multi-attribute decision making. Omega 29(5):405–415
3. Jovanovic M, Afgan N, Radovanovic P, Stevanovic V (2000) Sustainable development of the Belgrade energy system. Energy 34:532–539
4. Afgan NH, Carvalho MG (2002) Multi-criteria assessment of new and renewable energy power plants. Energy 27:739–755
5. Afgan NH, Carvalho MG (2004) Sustainability assessment of hydrogen energy systems. Int J Hydrogen Energ 29:1327–1342
6. Afgan NH, Carvalho MG (2008) Sustainability assessment of a hybrid energy system. Energ Policy 36:2903–2910
7. Liposcak M, Afgan NH, Duic N, daGraca Carvalho M (2006) Sustainability assessment of cogeneration sector development in Croatia. Energy 31:2276–2284
8. Pilavachi PA, Roumpeas CP, Minett S, Afgan NH (2006) Multi-criteria evaluation for CHP system options. Energ Convers Manage 47:3519–3529
9. Begic F, Afgan NH (2007) Sustainability assessment tool for the decision making in selection of energy system-Bosnian case. Energy 32:1979–1985

10. Afgan NH, Pilavachi PA, Carvalho MG (2007) Multi-criteria evaluation of natural gas resources. Energ Policy 35:704–713
11. Patlitzianas KD, Ntotas K, Doukas H, Psarras J (2007) Assessing the renewable energy producers' environment in EU accession member states. Energ Convers Manage 48:890–897
12. Renn O (2003) Social assessment of waste energy utilization scenarios. Energy 28:1345–1357

Chapter 5
PROMETHEE

PROMETHEE (Preference Ranking Organization Method for Enrichment Evaluation) is a multi-criteria decision-making method developed by Brans et al [1]. By 1994, the method has been extended to encompass six ranking formats: PROMETHEE I (partial ranking), PROMETHEE II (complete ranking), PROMETHEE III (ranking based on intervals), PROMETHEE IV (continuous case), PROMETHEE V (net flows and integer linear programming) and PROMETHEE VI (representation of human brain). The method uses the outranking methodology to rank the alternatives combined with the ease of use and decreased complexity. Based on extensions of the notion of criterion, the method is well adapted to problem where a finite number of alternative actions are to be ranked considering several criteria.

5.1 PROMETHEE I and II

The PROMETHEE method is implemented in five steps. In the first step, a preference function showing the preference of the decision-maker for an action a with regard to another action b, will be defined separately. The second step concerns the comparison of the suggested alternatives in pairs to the preference function. As a third step, the outcomes of these comparison are presented in an evaluation matrix as the estimated values of every criterion for every alternative. The ranking is realized in the two final steps: the fourth step includes the PROMETHEE I method application for partial ranking and afterward the fifth step includes PROMETHEE II method for complete ranking of the alternatives.

Step 1. Define the Preference function

Given the preference of the decision-maker for an action a with regard to b of a set of possible actions K, the preference function, which is a function of the difference between two alternatives for any criterion, will be defined separately for each criterion; its value will be between 0 and 1. The smaller the function, the greater the indifference of the decision-maker; the closer to 1, the greater his preference.

J. R. San Cristóbal Mateo, *Multi-Criteria Analysis in the Renewable Energy Industry*, 23
Green Energy and Technology, DOI: 10.1007/978-1-4471-2346-0_5,
© Springer-Verlag London Limited 2012

Fig. 5.1 Type I criterion

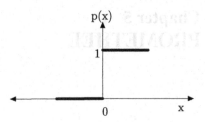

In case of strict preference, the preference function will be 1. The associated preference function $P(a, b)$ of a with regard to b will be defined as [2]:

$$P(a, b) = \begin{cases} 0, & \text{for } f(a) \le f(b) \\ p[f(a), f(b)], & \text{for } f(a) > f(b) \end{cases} \qquad (5.1)$$

For concrete cases, it seems reasonable to choose $p(\cdot)$ functions of the following type

$$p[f(a), f(b)] = p[f(a) - f(b)] \qquad (5.2)$$

depending on the difference between the values $f(a)$ and $f(b)$. Six types of functions cover most of the cases occurring in practical applications, namely usual criterion, quasi-criterion, criterion with lineal preference, level criterion, criterion with linear preference and indifference area and Gaussian criterion. In order to indicate clearly the areas of indifference in the neighbourhood of $f(a)$ and $f(b)$, we write:

$$x = f(a) - f(b) \qquad (5.3)$$

and the function $H(x)$ is graphically represented so that:

$$H(x) = \begin{cases} P(a, b), \ x \ge 0 \\ P(b, a), \ x \le 0 \end{cases} \qquad (5.4)$$

1. Type I: Usual Criterion.
   Let $p(x)$ be

$$p(x) = \begin{cases} 0, & \text{for } \forall x \le 0 \\ 1, & \text{for } \forall x > 0 \end{cases} \qquad (5.5)$$

In this case, there is indifference between a and b only when $f(a) = f(b)$. As soon as these values are different the decision-maker has a strict preference for the action having the greatest value. His preference function equals 1 and $H(x)$ is given in Fig. 5.1. If the decision-maker identifies the criterion $f(\cdot)$ as being of type I, no particular parameter has to be defined.

Fig. 5.2 Type II criterion

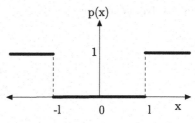

Fig. 5.3 Type III criterion

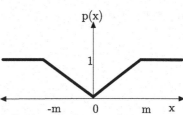

2. Type II: Quasi-Criterion.
   In this case,

$$p(x) = \begin{cases} 0, & \text{for } x \leq l \\ 1, & \text{for } x > l \end{cases} \tag{5.6}$$

and for the particular criterion, a and b are indifferent as long as the difference between $f(a)$ and $f(b)$ does not exceed l; if not, the preference becomes strict. When the decision-maker identifies the criterion $f(\cdot)$ as being of Type II, only the parameter l has to be defined. The function $H(x)$ is given in Fig. 5.2.

3. Type III: Criterion with Linear Preference
   Let $p(x)$ be

$$p(x) = \begin{cases} \frac{x}{m}, & \text{for } x \leq m \\ 1, & \text{for } x \geq m \end{cases} \tag{5.7}$$

Such an extension of the notion of criterion allows the decision-maker to prefer progressively a to b for progressively larger deviations between $f(a)$ and $f(b)$. In this case $H(x)$ is given in Fig. 5.3. The intensity of preference increases linearly until this deviation equals m, after this value the preference is strict. If the decision-maker considers that a particular criterion is of Type III, he has to define only the value m from which strict preference is considered.

4. Type IV: Level-Criterion

$$p(x) = \begin{cases} 0, & \text{for } x \leq q \\ \frac{1}{2}, & \text{for } q < x \leq q + p \\ 1, & \text{for } x > q + p \end{cases} \tag{5.8}$$

In this case, a and b are considered as indifferent when the deviation between $f(a)$ and $f(b)$ does not exceed q. Between q and $q + p$ the preference is weak

Fig. 5.4 Type IV criterion

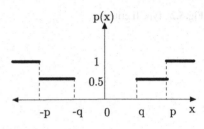

Fig. 5.5 Type V criterion

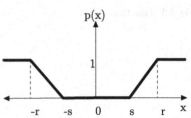

(1/2), after this value the preference becomes strict. $H(x)$ has the form depicted in Fig. 5.4. The decision-maker can easily fix q and p when he feels that the particular criterion $f(\cdot)$ is of Type IV. Criteria with more than two levels can also be considered, for instance, when several given norms seem relevant.

5. Type V: Criterion with Linear preference and Indifference Area.
   This time we consider for $p(x)$:

$$
p(x) = \begin{cases} 0, & \text{for } x \leq s \\ \frac{x-s}{r}, & \text{for } s \leq x \leq s+r \\ 1, & \text{for } x \geq s+r \end{cases} \tag{5.9}
$$

In this case the decision-maker considers that a and b are completely indifferent as long as the deviation between $f(a)$ and $f(b)$ does not exceed s. Above this value the preference grows progressively until this deviation equals $s+r$. $H(x)$ is then given by Fig. 5.5. Two parameters, s and r, have to be defined when a particular criterion has to be identified as being of this type.

6. Type VI: Gaussian Criterion
   Let $p(x)$ be:

$$
p(x) = \begin{cases} 0, & \text{for } x \leq 0 \\ 1 - e^{-\frac{x^2}{2\sigma^2}}, & \text{for } x \geq 0 \end{cases} \tag{5.10}
$$

If a particular criterion is of the Gaussian type, the preference of the decision-maker still grows with the deviation x. The value of σ is the distance between the origin and the point of inflexion of the curve. $H(x)$ then is given by Fig. 5.6. In this particular case only the value of σ has to be defined by the decision-maker.

Fig. 5.6 Type VI criterion

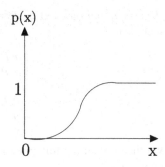

Step 2. Calculate the Preference index

Suppose every criterion has been identified as being one of the six types considered so that the preference functions $P_h(a, b)$ have been defined for each $h = 1, 2, \ldots, k$. Next, for each couple of actions $a, b \in K$, we first define a preference index for a with regard to b over all the criteria. Let

$$\pi(a, b) = \frac{1}{k} \sum_{h=1}^{K} P_h(a, b) \tag{5.11}$$

be such a preference index, which gives a measure of the preference of a over b for all the criteria. The closer to 1, the greater the preference.

Step 3. Construct the valued outranking graph

The values obtained in Step 2 determine the valued outranking graph, the nodes of which are the actions of K, so that for all $a, b \in K$, the arc (a, b) has the value $\pi(a, b)$. Let us define, for each node in this valued outranking graph, the outgoing flow:

$$\phi^+(a) = \sum_{x \in K} \pi(a, x) \tag{5.12}$$

and the incoming flow:

$$\phi^-(a) = \sum_{x \in K} \pi(x, a) \tag{5.13}$$

The larger $\phi^+(a)$, the more a dominates the other actions of K. The smaller $\phi^-(a)$, the less a is dominated.

Step 4. Ranking the actions by a partial preorder

If the decision-maker wants to rank the actions of K from the best to the weakest one, the problem consists in using the outranking graph to build a total preorder on

K, or a partial one. Let us define the two total preorders (P^+, I^+) and (P^-, I^-) such that

$$\begin{cases} a \ P^+ \ b & \text{if } \phi^+(a) > \phi^+(b) \\ a \ P^- \ b & \text{if } \phi^-(a) < \phi^-(b) \end{cases} \tag{5.14}$$

$$\begin{cases} a \ I^+ \ b & \text{if } \phi^+(a) = \phi^+(b) \\ a \ I^- \ b & \text{if } \phi^-(a) = \phi^-(b) \end{cases} \tag{5.15}$$

We then obtain the following partial preorder $(P^{(1)}, I^{(1)}, R)$ by considering their intersection:

$$\begin{cases} a \text{ outranks } b \ (a \ P^{(1)} \ b) & \text{if} \begin{cases} a \ P^+ \ b \text{ and } a \ P^- \ b \\ a \ P^+ \ b \text{ and } a \ I^- \ b \\ a \ I^+ \ b \text{ and } a \ P^+ \ b \end{cases} \\ a \text{ is indifferent to } b \ (a \ I^{(1)} \ b) & \text{if } a \ I^+ \ b \text{ and } a \ I^- \ b \\ a \text{ and } b \text{ are incomparable } (a \ R \ b) & \text{otherwise} \end{cases} \tag{5.16}$$

This is the PROMETHEE I partial relation. It offers the decision-maker a graph in which some actions are comparable, while some others are not.

Step 5. Ranking the actions by a total preorder

Suppose a total preorder (complete ranking without incomparabilities) has been requested by a decision-maker. We then consider for each criterion $a \in K$ the net flow:

$$\phi(a) = \phi^+(a) - \phi^-(a) \tag{5.17}$$

which can easily be used for ranking the actions:

$$\begin{cases} a \text{ outranks } b \ (a \ P^{(2)} \ b) & \text{if } \phi(a) > \phi(b) \\ a \text{ is indifferent to } b \ (a \ I^{(2)} \ b) & \text{if } \phi(a) = \phi(b) \end{cases} \tag{5.18}$$

This is the PROMETHEE II complete relation. All the actions of K are now completely ranked but this relation is poorer in information and less realistic because of the balancing effects between outgoing and incoming flows.

5.2 Application

Let us consider the multi-criteria problem proposed by Goumas and Lygerou [3] consisting in the selection of four alternative exploitation scenarios for the low enthalpy geothermal field in a rural community located in Northern Greece. Each exploitation scenario includes a number of possible uses such as: greenhouse heating, subsoil heating, drying of agricultural products and water heating for fish farming. For

Table 5.1 Alternatives for a geothermal field [3]

Alternatives	NPV ($10^6 GRD$)	Jobs (N)	Energy used (10^{12} cal/year)	Risk
	Type II	Type III	Type V	Type IV
	l = 100	m = 10	s = 1	q = 1
			r = 3	p = 3
A_1	373	44	22.4	3
A_2	3706	99	51.0	6
A_3	3809	100	49.3	8
A_4	3860	104	47.5	9
weight	0.30	0.30	0.10	0.30

more details on the formation of alternative exploitation schemes see Goumas and Lygerou [3], four criteria have been selected for the evaluation: the Net Present Value of the investment (NPV in Greek Drachmes), the creation of new jobs (Jobs), the Energy consumed which indicates the degree of exploitation of the resource and the Risk associated either with new products and/or the application of new technology or with uncertainties for future changes in the economic parameters. The index is expressed in a scale of 1 (no risk) to 20 (very risky). The criteria NPV, and Jobs have to be maximized whereas Energy used and Risk have to be maximized.

The data for the evaluation, the type of each criterion and the the values of the corresponding parameters are shown in Table 5.1. In Goumas and Lygerou [3], the generalized criterion with linear preference was applied in all criteria. To show the use of different criteria, the following types of criterion have been considered:

1. Net Present Value. Type of criterion II: Quasi-criterion.

$$p(x) = \begin{cases} 0, & \text{for } x \le 100 \\ 1, & \text{for } x > 100 \end{cases}$$

   In this case two alternatives are indifferent as long as the difference between them regarding the investment required does not exceed 100 GRD. If not, the preference becomes strict.

2. Creation of new Jobs. Type of criterion III. Criterion with Linear Preference

$$p(x) = \begin{cases} \frac{x}{10}, & \text{for } x \le 10 \\ 1, & \text{for } x \ge 10 \end{cases}$$

   This criterion allows the decision-maker to prefer progressively alternative a to b for progressively larger deviations between $f(a)$ and $f(b)$. The intensity of preference increases linearly until this deviation equals 10 jobs. After this value the preference is strict.

3. Energy consumed. Type of criterion V: Criterion with Linear Preference and Indifference Area

Table 5.2 Values of
$\pi(A_i, A_k)$

	A_1	A_2	A_3	A_4
A_1	–	0.375	0.500	0.500
A_2	0.500	–	0.125	0.125
A_3	0.500	0.332	–	0
A_4	0.500	0.707	0.315	–

Table 5.3 Data for preorders P^+ and P^-

	A_1	A_2	A_3	A_4
$\phi^+(A_i)$	1.375	0.750	0.832	1.522
$\phi^-(A_k)$	1.500	1.414	0.940	0.625

$$p(x) = \begin{cases} 0, & \text{for } x \leq 1 \\ \frac{x-1}{3}, & \text{for } 1 \leq x \leq 4 \\ 1, & \text{for } x \geq 4 \end{cases}$$

In this case the decision-maker considers that two alternatives are completely indifferent as long as the deviation between them does not exceed $1 * 10^{12}$ cal/year. Above this value the preference grows progressively until this deviation equals $4 * 10^{12}$ cal/year hours.

4. Risk. Type of criterion IV: Level Criterion

$$p(x) = \begin{cases} 0, & \text{for } x \leq 1 \\ \frac{1}{2}, & \text{for } 1 < x \leq 4 \\ 1, & \text{for } x > 4 \end{cases}$$

In this case, two alternatives are considered as indifferent when the deviation between them does not exceed 1. Between 1 and 4 the preference is weak, and after this value the preference becomes strict.

Using Eqs. 5.1, 5.3 and 5.11 in the right way according to whether the criterion has to be minimized or maximized, we obtain the preference index for each couple of alternatives. Thus, for alternatives (A_1) and (A_2) the preference index will be:

$$\phi(A_1, A_2) = \frac{1}{4}[P_1(A_1, A_2) + P_2(A_1, A_2) + P_3(A_1, A_2) + P_4(A_1, A_2)]$$
$$= \frac{1}{4}[0 + 0 + 1 + \frac{1}{2}] = 0.375$$

Table 5.2 shows the values $\pi(A_i, A_k)$ for every couple of actions. These values determine the outranking graph, each arc (A_i, A_k) having the value $\phi(A_i, A_k)$.

Let us first suppose that a partial relation would be useful to the decision-maker. We therefore apply the PROMETHEE I technique. According to Eqs. 5.12 and 5.13 we complete Table 5.3 with the outgoing and the incoming flow.

Next, we obtain the preorders P^+ and P^- from Table 5.3, the intersection of which is:

$$A_4 P^{(1)} A_1, \quad A_4 P^{(1)} A_2, \quad A_4 P^{(1)} A_3, \quad A_3 P^{(1)} A_2$$

Table 5.4 Net flows

	A_1	A_2	A_3	A_4
$\phi(A_i)$	−0.125	−0.644	−0.108	0.897

Supposing now that the decision-maker requests a total preorder, we can then use PROMETHEE II. According to Eq. 5.17 we calculate the net flows shown in Table 5.4. From Table 5.4, alternative A_4 is considered the best ranked alternative.

5.3 Concluding Remarks

PROMETHEE methods are well adapted to problems where a finite number of alternative actions are to be ranked considering several criteria. The methods are rather popular in the world of outranking methods and one of the reasons for this popularity is the existence of user friendly softwares such as PROMCALC. However, when applying the method several considerations must be taken into account. For example, the decision-maker can express his preferences between two actions on all criteria and the importance he attaches to the criteria on a ration scale, for all criteria the difference between evaluations must be meaningful.

Mladineo et al [4] used PROMETHEE method to deal with the problem of small hydropower construction in a wide area covering many regions in order to make the units costs as low as possible. Georgopoulou et al. [5] designed and implemented a decision support system by means of PROMETHEE II for sustainable renewable energy exploitation in Greece. Diakoulaki and Karangelis [6], Goumas et al. [7] and Goumas and Lygerou [3] applied the method to geothermal energy projects and Haralambopoulos and Polatidis [8] to the exploitation of a geothermal resource in the island of Chios (Greece). Pohekar and Ramachandran [9] used the method for promoting parabolic solar cooker in India. Madlener et al. [10] explored future renewable energy options in Austria and Tsoutos et al. [11] the sustainable energy planning on the island of Crete.

References

1. Brans JP, Vincke PH, Mareschal B (1986) How to select and how to rank projects: the PROMETHEE method. Eur J Oper Res 24:228–238
2. Brans JP, Vincke PH (1985) A preference ranking organization method: the PROMETHEE method for multiple-criteria decision-making. Manage Sci 31(6):647–656
3. Goumas M, Lygerou V (2000) An extension of the PROMETHEE method for decision making in fuzzy environment: ranking of alternative energy exploitation projects. Eur J Oper Res 123:606–613
4. Mladineo N, Margeta J, Brans JP, Mareschal B (1987) Multicriteria ranking of alternative locations for small scale hydro plants. Eur J Oper Res 31:215–222

5. Georgopoulou E, Sarafidis Y, Diakoulaki D (1998) Design and implementation of a group DSS for sustaining renewable energies exploitation. Eur J Oper Res 109(2):483–500
6. Diakoulaki D, Karangelis F (2007) Multi-criteria decision analysis and cost-benefit analysis of alternative scenarios for the power generation sector in Greece. Renew Sust Energ Rev 11:716–727
7. Goumas MG, Lygerou VA, Papayannakis LE (1999) Computational methods for planning and evaluating geothermal energy projects. Energ Policy 27:147–154
8. Haralambopoulos A, Polatidis H (2003) Renewable energy projects: structuring a multi-criteria group decision-making framework. Renew Energ 28:961–973
9. Pohekar SD, Ramachandran M (2004) Multi-criteria evaluation of cooking energy alternatives for promoting parabolic solar cooker in India. Renew Energ 29:1449–1460
10. Madlener R, Kowalski K, Stagl S (2007) New ways for the integrated appraisal of national energy scenarios: the case of renewable energy use in Austria. Energ Policy 35:6060–6074
11. Tsoutos T, Drandaki M, Frantzeeskaki N et al (2009) Sustainable energy planning by using multi-criteria analysis application in the island of Crete. Energ Policy 37:1587–1600

Chapter 6
ELECTRE

The methods belonging to the ELECTRE family (ELimination Et Choix Traduisant la REalité) devised by Roy [1] grew from the idea that rigorous mathematical axioms are unsuited to describe a complex reality such as the one of a contradiction-laden decision process. ELECTRE methods are based on the partial aggregation of preferences, which is based on the principle that the decision-maker is not perfectly rational and can therefore express, for each pair of actions (a and b), not only the preference (P) or indifference (I) toward one action or the other, but also a less marked preference labeled 'weak' (Q) or a situation of incomparability (R).

ELECTRE methods are not compensatory, a high degree of dissatisfaction regarding the performance of one alternative compared to another for one specific criterion may not necessarily be compensated for by good performance with regard to the other remaining criteria. It may therefore happen that the alternatives differ so widely from one another that the decision-maker is unable to express a preference as they are not comparable.

6.1 ELECTRE III

The distinctive characteristic of ELECTRE III is its ability to handle a data-set affected by a high degree of uncertainty. It also has the advantage of being less sensitive to any changes in data, which explains why ELECTRE III is very well-suited to environmental applications and it also involves some aspects neglected by other methods [2]. The method is capable of handling discrete criteria both quantitative and qualitative in nature.

The evaluation procedures of the ELECTRE III method encompass the establishment of the threshold function, disclosure of concord index and discord index, outranking degree and the ranking of alternatives [3].

J. R. San Cristóbal Mateo, *Multi-Criteria Analysis in the Renewable Energy Industry*,
Green Energy and Technology, DOI: 10.1007/978-1-4471-2346-0_6,
© Springer-Verlag London Limited 2012

Step 1. Establishment of the threshold function

Let q_j and p_j represent the indifference threshold and preference threshold, respectively, with regard to the jth criterion. Indifference and preference thresholds characterize the acceptance of imprecision in the judgment by considering two entities as indifferent when their individual performance in each criterion j differs less than a specified amount q_j. In particular, q_j indicates the minimum boundary of uncertainty associated with the performed calculations, while p_j can be considered as the maximum boundary of error, connected to the performed calculations. Moreover, the transition from indifference to preference is made gradual, changing linearly from q_j to p_j. Threshold functions are linked to the idea of a pseudo-criterion. A criterion g_j is a pseudo-criterion if there exists two threshold functions $q(g)$ and $p(g)$ that represent the indifference and preference threshold, respectively, so that for $g(a) \geq g(b)$ there are situations in-between in which a weak preference for option a over b denoted aQb applies as follows [3].

1. $g(a) > g(b) + p(g(b)) \Leftrightarrow aPb$

2. $g(b) + q(g(b)) < g(a) \leq g(b) + p(g(b)) \Leftrightarrow aQb$

3. $g(b) \leq g(a) \leq g(b) + q(g(b)) \Leftrightarrow aIb$

where P refers to strong preference, Q weak preference, I indifference and $g(a)$ and $g(b)$ are the criteria value of alternative a and b, respectively.

The establishment of a threshold function has to satisfy the subsequent constraint equations:

1. $g(a) > g(b) \rightarrow g(a) + q(g(a)) \geq g(b) + q(g(b))$
$$g(a) + p(g(a)) \geq g(b) + p(g(b))$$

2. for all criteria, $p(g) \geq q(g)$

$p_j(g_j(a))$ and $q_j(g_j(a))$ can be assessed according to Roy's formula [4]:

$$p_j(g_j(a)) = \alpha_p + \beta_p g_j(a)$$

$$q_j(g_j(a)) = \alpha_q + \beta_q g_j(a)$$

where $g_j(a)$ is the evaluation value of alternative a on criterion j. The threshold values may be constant (α is determined and $\beta = 0$) or proportional ($\beta > 0$).

Step 2. Concordance index and Discordance index

A Concordance index $C(a,b)$ is computed for each pair of alternatives:

$$C(a, b) = \frac{1}{k} \sum_{j=1}^{n} k_j c_j(a, b) \qquad (6.1)$$

where k_j is the weight of criterion j and $c_j(a, b)$ is the outranking degree of alternative a and alternative b under criterion j, calculated as follows:

$$c_j(a, b) = \begin{cases} 1 & \text{if } g_j(a) + q_j \geq g_j(b) \\ 0 & \text{if } g_j(a) + p_j \leq g_j(b) \\ \frac{p_j + g_j(a) - g_j(b)}{p_j - q_j} & \text{otherwise} \end{cases} \qquad (6.2)$$

A discordance index $d_j(a, b)$ for each criterion is then defined as:

$$d_j(a, b) = \begin{cases} 0 & \text{if } g_j(a) + p_j \geq g_j(b) \\ 1 & \text{if } g_j(a) + v_j \leq g_j(b) \\ \frac{g_j(b) - g_j(a) - p_j}{v_j - p_j} & \text{otherwise} \end{cases} \qquad (6.3)$$

where v_j is the veto threshold value defined for each criterion j. The veto threshold value is the limit value of the difference $g_j(b) - g_j(a)$, over which it is reasonable to reject the hypothesis of outranking of a over b, with regard to the considered criterion. The veto threshold is aimed at capturing situations in which very bad scores in any criterion should prevent an entity of being classified in the best category, or if these bad scores should force it to be classified in the worst category independent of having very good scores in all other criteria.

Step 3. Outranking degree

The next step is to combine the concordance and discordance measure for each pairs of alternatives to produce a measure of the degree of outranking; that is, a credibility matrix which assesses the strength of the assertion a is at least as good as b. The credibility degree for each pair of alternatives is defined by $S(a, b)$:

$$S(a, b) = \begin{cases} C(a, b), & \text{if } d_j(a, b) \leq C(a, b) \forall j \\ C(a, b) * \prod_{j \in J(a,b)} & \text{otherwise} \\ \times (1 - d_j(a, b))/(1 - c_j(a, b)), \end{cases} \qquad (6.4)$$

where $J(a,b)$ is the set of criteria for which $d_j(a, b) > C(a, b)$

This formula assumes that if the strength of the concordance exceeds that of the discordance, then the concordance value should be modified. Otherwise, we are force to question the assertion that aSb and modify $C(a, b)$ according to the above equation.

Step 4. Ranking of alternatives

To obtain a complete ranking of the alternatives, the normal ranking method of ELEC-TRE III uses a structured algorithm via two intermediate ranking procedures: one descending where alternatives are classified from the best to the worst (descending distillation) and the other ascending from the worst to the best (ascending distillation). The partial rankings obtained take into account indifference but not incomparability and they are therefore intertwined in order to obtain the final ranking in which incomparability is also taken into account. Following Li and Wang [5] the weakness of the normal ranking of ELECTRE III is that it requires an additional threshold to be introduced and the ranking of the alternatives depends on the size of this threshold for which there exists no 'correct' value. Aiming at the ranking problems in ELECTRE III, Li and Wang [5] present a new ranking method by introducing three concepts, the concordance credibility degree, the discordance credibility degree and the net credibility degree.

1. The concordance credibility degree is defined by

$$\Phi^+(x_i) = \sum_{x_j \in X} S(x_i, x_j), \ \forall x_i \in X \tag{6.5}$$

The concordance credibility degree is the measure of the outranking character of x_i (how x_i dominates all the other alternatives of X).
2. The discordance credibility degree is defined by

$$\Phi^-(x_i) = \sum_{x_j \in X} S(x_j, x_i), \ \forall x_i \in X \tag{6.6}$$

The discordance credibility degree gives the outranked character of x_i (how x_i is dominated by all the other alternatives of X).
3. The net credibility degree is defined by

$$\phi(x_i) = \Phi^+(x_i) - \Phi^-(x_i), \ \forall x_i \in X \tag{6.7}$$

The net credibility degree represents a value function, where a higher value reflects a higher attractiveness of alternative x_i.

Next, all the alternatives can be completely ranked by the net credibility degree.

6.2 ELECTRE IV

The basic difference between ELECTRE III and IV is that no weights for the criteria are introduced in ELECTRE IV, which does not mean that the weights of the criteria are considered to be equal. The ELECTRE IV method uses the following notation [3]:

- $m_p(a, b)$: the number of criteria for which alternative a is strictly preferred to alternative b.
- $m_q(a, b)$: the number of criteria for which alternative a is weakly preferred to alternative b.
- $m_i(a, b)$: the number of criteria for which alternative a is considered indifferent to alternative b but such that a has a better criterion value than b.
- $m_o(a, b) = m_o(b, a)$: the number of equal criterion values of alternative a and alternative b.

Table 6.1 Evaluation matrix

Alternatives	Power (Mw)	NPV(Euros $* 10^6$)	Maturity (1–5)	Emissions (tCO_2/y)	Jobs
A_1 Energetic cultiv.	245	7,500,000	2	2,250,000	15
A_2 Forest and agric. wastes	260	8,650,000	5	2,650,000	22
A_3 Farming indust. wastes	100	9,300,000	4	2,460,000	18
A_4 Forest indust. wastes	230	9,650,000	3	3,100,000	25

The outranking relations in the ELECTRE IV method are the following:

1. Quasi-dominance S_q
   $aS_qb \Leftrightarrow m_p(b, a) + m_q(a, b) = 0$ and $m_i(b, a) \leq 1 + m_i(a, b) + m_q(a, b) + m_p(a, b)$
2. Canonic-dominance S_c
   $aS_cb \Leftrightarrow m_p(b, a) = 0$ and $m_q(b, a) \leq m_p(a, b)$ and $m_q(b, a) + m_i(b, a) \leq 1 + m_i(a, b) + m_q(a, b) + m_p(a, b)$
3. Pseudo-dominance S_p
   $aS_pb \Leftrightarrow m_p(b, a) = 0$ and $m_q(b, a) \leq m_q(a, b) + m_p(a, b)$
4. Veto-dominance S_v
   $aS_vb \Leftrightarrow m_p(b, a) = 0$ if $m_p(b, a) = 1$ and no $bPV_ja, \forall j$ and $m_p(a, b) \geq \frac{m}{2}$ in which $bPV_ja \Leftrightarrow g_j(b) \geq g_j(a) + v_j(g_j(a))$

The partial preorder is constructed in a similar way as in ELECTRE III.

6.3 Application

The selection among four different Biomass plants for electric generation is considered in order to show the use of ELECTRE III method. The evaluation matrix is shown in Table 6.1. The alternatives will be evaluated according to the following criteria: Power (Mw), Net Present Value (NPV), Maturity (M), tons of emissions of CO_2 avoided per year (tCO_2/y) and number of Jobs created (J).

Once the criteria has been defined, it is necessary to assign weights to these criteria. In ELECTRE methods, weights are perceived as true coefficients of importance assigned to the criteria, and not just as technical devices for translating the performance in the criteria considered into a common value measure. Therefore, they are scale-independent, that is, they are not linked to the scales in which each criterion is measured, thus making them easier to be specified by managers [6]. The most used technique when determining the weights in ELECTRE methods is the Sismos approach [7]. Table 6.2 shows the weighting procedure, according to this technique,

Table 6.2 Weighting procedure according to Simos approach

Ranking[a]	Criteria	N_r	Weight W	Average weight $\bar{W} = \frac{\sum W}{N_r}$	Relative weight $W_r = \frac{\bar{W}}{\sum \bar{W}}$	Test
1	c, e	2	1, 2	$\frac{(1+2)}{2} = 1.5$	10	$2*10 = 20$
2	a, b	2	3, 4	$\frac{(3+4)}{2} = 3.5$	23	$2*23 = 46$
3	d	1	5	1	33.4	$1*34 = 34$
			15^b			100

[a] from the best to the worst criterion [b] without the weights in brackets, where N_r is the number of criteria in each place, and the relative weight is rounded off to percent

Table 6.3 Weights and threshold values

	Power	NPV	Maturity	Emissions	Jobs
Weights	0.23	0.23	0.10	0.34	0.10
Indifference threshold (q)	15	300,000	1	250,000	3
Preference threshold (p)	30	500,000	2	400,000	5
Veto threshold (v)	70	750,000	3	500,000	

based on a first ranking of criteria and a subsequent assignment of weights depending on each criterion rank.

Once the problem has been defined, the alternatives generated and the weights assigned, threshold values must be considered. Procedures for choosing appropriate threshold values are addressed by Rogers and Bruen [8]. In our case, we have considered the indifference, preference and veto threshold shown in Table 6.3.

Next, the concordance index is computed. The concordance calculations for alternatives A_1 and A_2 are as follows:

$$c_1(A_3, A_4) = 0 \quad \text{since } 100 + 15 \leq 230$$
$$\text{and } 100 + 30 \leq 230$$
$$c_2(A_3, A_4) = 0.75 \quad \text{since } 9,300,000 + 300,000 \leq 9,650,000$$
$$\text{and } 9,300,000 + 500,000 \geq 9,650,000$$
$$\text{then } \frac{500,000 + 9,300,000 - 9,650,000}{500,000 - 300,000} = 0.75$$
$$c_3(A_3, A_4) = 1 \quad \text{since } 4 + 1 \geq 3$$
$$c_4(A_3, A_4) = 0 \quad \text{since } 42,460,000 + 250,000 \leq 3,100,000$$
$$\text{and } 2,460,000 + 400,000 \leq 3,100,000$$
$$c_5(A_3, A_4) = 0 \quad \text{since } 18 + 3 \leq 25$$
$$\text{and } 18 + 3 \leq 25$$

Therefore $C(A_3, A_4) = 0.23 * 0.75 + 0.10 * 1 = 0.27$.

The value of 0.27 measures the strength of the assertion that A_3 is at least as good as A_4. Table 6.4 presents the complete concordance matrix.

In the concordance matrix we have a measure of the extent to which we are in harmony with the assertion that alternative a is at least as good as alternative b. Next, to calculate if there is any discordance associated with the assertion aSb the veto threshold is used. This veto threshold allows for the possibility of aSb to be refused totally if, for any one criterion j, $g_j(a) + v_j < g_j(b)$.

Consider criterion c_2, with a veto threshold of 750,000. Comparing A_1 and A_4, it is clear that $7,500,000 + 500,000 < 800,000$ or $g_j(a) + v_j < g_j(b)$.

Therefore, the discordance index $d_1(A_1, A_4) = 1$. Unlike concordance, no aggregation over criteria takes place; one discordant criterion is sufficient to discard outranking.

There now exists a concordance and a discordance measure for each pair of alternatives. The next step is to combine these two measures to produce a measure

Table 6.4 Concordance matrix

	A_1	A_2	A_3	A_4
A_1	1	0.23	0.67	0.33
A_2	1	1	0.77	0.43
A_3	0.77	0.67	1	0.27
A_4	1	0.67	1	1

Table 6.5 Credibility matrix

	A_1	A_2	A_3	A_4
A_1	1	0	0	0
A_2	1	1	0	0
A_3	0	0	1	1
A_4	1	0.67	1	1

Table 6.6 Concordance, discordance and net credibility matrix

Alternative	ϕ^+	ϕ^-	ϕ
A_1	1	3	-2
A_2	2	1.67	0.33
A_3	1	2	-1
A_4	3.67	1	2.67

of the degree of outranking, that is, a credibility matrix which assesses the strength of the assertion that '*a* is at least as good as *b*'. If the discordance is 1.00 for any criterion, then we have no confidence that aSb and $S(a, b) = 0.00$. The credibility matrix for our multi-criteria problem is shown in Table 6.5.

The next step in the model is to produce a ranking of alternatives from the credibility matrix. Following Li and Wang [5] the concordance, discordance and net credibility degree are calculated by using Eqs. 6.5, 6.6 and 6.7 respectively and shown in Table 6.6. From Table 6.6, we may conclude that alternative A_4, Forest industrial wastes, is the best ranked alternative.

6.4 Concluding Remarks

Applications of ELECTRE methods can be seen in Barda et al. [9] for the location of thermal power plants, Georgopoulou et al. [10] for the renewable energy planning in a Greek island. Beccali et al. [11] for the diffusion of RE technologies in the island of Sardina (Italy), Papadopoulos and Karagiannidis [12] for the optimization of decentralized energy systems, Madlener et al. [6] to assess the performance of agricultural biogass plants and Caravallo [2] to assess a selection of production processes of thin-film solar technology.

References

1. Roy B (1968) Classement et choix en présence de critères multiples (le méthode ELECTRE). RIRO 8:57–75
2. Cavallaro F (2010) A comparative assessment of thin-film photovoltaic production processes using the ELECTRE III method. Energ Policy 38:463–474
3. Hokkanen J, Salminen P (1997) ELECTRE III and IV decision aids in an environmental problem. J MultiCrit Decis Anal 6:215–226
4. Skalka JM, Bouyssou D, Vallée D (1992). ELECTRE III et IV. Aspects méthodologiques et guide d'utilisation, Document du LAMSADE No. 25, Laboratoire d'Analyse et Modélisation de Systémes pour l'Aide á la Décision, Université de Paris-Dauphine.
5. Li HF, Wang JJ (2007). An improved ranking method for ELECTRE III. doi:1-4244-1312-5/07/25.00
6. Madlener R, Antunes CH, Dias LC (2009) Assessing the performance of biogas plants with multi-criteria and data envelopment analysis. Eur J Oper Res 197:1084–1094
7. Simos J (1990) Evaluer l'impact sur l'environnement. Une approche originale par l'analyse multicrite're de ne'gotiation. Presse Polytechniques et Universitaires Romandes, Lausanne
8. Rogers M, Bruen M (1998) A new system for weighting environmental criteria for use within ELECTRE III. Eur J Oper Res 107:552–563
9. Barda OH, Dupluis J, Lencoini P (1990) Multicriteria location of thermal power plants. Eur J Oper Res 45:332–346
10. Georgopoulou E, Lalas D, Papagiannakis L (1997) A multicriteria decision aid approach for energy planning problems: the case of renewable energy option. Eur J Oper Res 103:38–54
11. Beccali M, Cellura M, Mistretta M (2003) Decision-making in energy planning-application of the ELECTRE method at regional level for the diffusion of renewable energy technology. Renewable Energy 28(13):2063–2087
12. Papadopoulos A, Karagiannidis A (2008) Application of the multi-criteria analysis method ELECTRE III for the optimisation of decentralised energy systems. Omega 36:766–776

Chapter 7
TOPSIS

The principle behind the Technique for Order Preference by Similarity to Ideal Solutions (TOPSIS) is simple. Ideal and a negative-ideal solutions are formed. The ideal solution is formed as a composite of the best performance value exhibited by any alternative for each attribute and the negative-ideal solution is the composite of the worst performance values. The chosen alternative should be as close to the ideal solution as possible and as far from the negative-ideal solution as possible. In this chapter, the method is applied to the selection of a renewable project for electric generation.

7.1 TOPSIS

The Technique for Order Preference by Similarity to Ideal Solutions (TOPSIS) method is developed by Hwang and Yoon [1] as an alternative to ELECTRE. The basic principle is that the selected alternative should have the shortest distance from the ideal solution and the farthest distance from the negative-ideal solution in geometrical sense. The method assumes that each attribute has a monotonically increasing or decreasing utility. This makes it easy to locate the ideal and negative-ideal solutions. Thus, the preference order of alternatives is yielded through comparing the Euclidean distances. The TOPSIS method consists of the following steps:

Step 1. Calculate the normalized decision matrix

A decision matrix of the m alternatives and n criteria is formulated first. Then the normalized value r_{ij} is calculated as

$$r_{ij} = \frac{f_{ij}}{\sqrt{\sum_{j=1}^{m} f_{ij}^2}} \qquad (7.1)$$

where f_{ij} is the value of the ith criterion function for the alternative A_j ($j = 1, \ldots, m$; $i = 1, \ldots, n$).

J. R. San Cristóbal Mateo, *Multi-Criteria Analysis in the Renewable Energy Industry*,   43
Green Energy and Technology, DOI: 10.1007/978-1-4471-2346-0_7,
© Springer-Verlag London Limited 2012

Step 2. Calculate the weighted normalized decision matrix

The weighted normalized value v_{ij} is calculated as:

$$v_{ij} = w_i r_{ij} \tag{7.2}$$

where w_i is the weight of the i criterion or attribute and $\sum_{i=1}^{n} w_i = 1$.

Step 3. Determine the ideal and negative-ideal solutions

For benefit criteria the decision-maker wants to have maximum values among the alternatives and for cost criteria he wants minimum values among alternatives. Then, the ideal solutions (A^*) and the negative-ideal solutions (A^-) will be:

$$A^* = \{v_1^*, \ldots, v_n^*\} = \{(\max_j v_{ij} | i \in I'), (\min_j v_{ij} | i \in I'')\} \tag{7.3}$$

$$A^- = \{v_1^-, \ldots, v_n^-\} = \{(\min_j v_{ij} | i \in I'), (\max_j v_{ij} | i \in I'')\} \tag{7.4}$$

where I' is associated with benefit criteria, and I'' is associated with cost criteria

Step 4. Calculate the separation measures

Using the n-dimensional Euclidean distance, the separation of each alternative from the ideal solution is given as:

$$D_j^* = \sqrt{\sum_{i=1}^{n} (v_{ij} - vi^*)^2} \tag{7.5}$$

Similarly, the separation from the negative-ideal solution is given as:

$$D_j^- = \sqrt{\sum_{i=1}^{n} (v_{ij} - vi^-)^2} \tag{7.6}$$

Step 5. Calculate the relative closeness to the ideal solution

The relative closeness of the alternative a_j with respect to A^* is defined as:

$$C_j^* = \frac{D_j^-}{(D_j^* + D_j^-)} \tag{7.7}$$

Step 6. Rank the preference order

Rank the alternatives, sorting by the value C_j^* in decreasing order. Propose as a solution the alternative which is the best ranked (maximum) by the measure.

Table 7.1 Alternatives for electric generation

Alternatives	
A_1	Wind power $P \leq 5Mw$
A_2	Wind power $5 \leq P \leq 10Mw$
A_3	Wind power $10 \leq P \leq 50Mw$
A_4	Hydroelectric $P \leq 10Mw$
A_5	Hydroelectric $10 \leq P \leq 25Mw$
A_6	Hydroelectric $25 \leq P \leq 50Mw$
A_7	Solar Thermo-electric $P \geq 10Mw$
A_8	Biomass (energetic cultivations) $P \leq 5Mw$
A_9	Biomass (forest and agricultural wastes) $P \leq 5Mw$
A_{10}	Biomass (farming industrial wastes) $P \leq 5Mw$
A_{11}	Biomass (forest industrial wastes) $P \leq 5Mw$
A_{12}	Biomass (Co-combustion in conventional central) $P \geq 50Mw$
A_{13}	Bio fuels $P \leq 2Mw$

Table 7.2 Criteria used to evaluate the alternatives

	Criteria	Unit	Weight
f_1	Power (P)	Kw	0.32
f_2	Investment ratio (IR)	$Euros/Kw$	0.09
f_3	Implement period (IP)	$Years$	0.03
f_4	Operating hours (OH)	$Hours/year$	0.12
f_5	Useful life (UL)	$Years$	0.13
f_6	Operation and maintenance costs (OM)	$Euros * 10^{-3}/Kwh$	0.04
f_7	Tons of CO_2 avoided (tCO_2/y)	$Tons$	0.27

7.2 Application

Let us consider in this section as a multi-criteria problem, the selection among 13 different RE alternatives for electric generation shown in Table 7.1. The designed systems will be evaluated according to the criteria shown in Table 7.2. The attributes considered are: Power (P), Investment Ratio (IR), Implementation Period (IP), Operating Hours (OH), Useful Life (UL), Operation and Maintenance costs (OM) and tons of emissions of CO_2 avoided per year (tCO_2/y). The weights have been estimated by AHP. The evaluation matrix showing the data considering the 13 alternative RE projects and 7 selection criteria are shown in Table 7.3.

Firstly, the normalized decision matrix is calculated using Eq. 7.1. Then, using Eq. 7.2, the construction of the weighted decision matrix, shown in Table 7.4, is carried out. Table 7.5 shows the ideal and negative-ideal solutions, determined from the weighted normalized decision matrix, taking into account that, for benefit criteria the decision-maker wants to have maximum values among the alternatives and for cost criteria, he wants minimum values among the alternatives. This is followed by the separation measures of each alternative from the ideal solution and from the negative-ideal solution calculated using Eqs. 7.5 and 7.6, respectively. Next, using

Table 7.3 Evaluation matrix

Criteria							
	P	IR	IP	OH	UL	OM	tCO_2/y
A_1	5,000	937	1	2,350	20	1.47	1,929,936
A_2	10,000	937	1	2,350	20	1.47	3,216,560
A_3	25,000	937	1	2,350	20	1.51	9,649,680
A_4	5,000	1,500	1.5	3,100	25	1.45	472,812
A_5	20,000	700	2	2,000	25	0.70	255,490
A_6	35,000	601	2.5	2,000	25	0.60	255,490
A_7	50,000	5,000	2	2,596	25	4.20	482,856
A_8	5,000	1,803	1	7,500	15	7.11	2,524,643
A_9	5,000	1.803	1	7,500	15	5.42	2,524,643
A_{10}	5,000	1,803	1	7,500	15	5.42	2,524,643
A_{11}	5,000	1,803	1	7,500	15	2.81	2,524,643
A_{12}	56,000	856	1	7,500	20	4.56	22,524,643
A_{13}	2,000	1,503	1	7,000	20	2.51	5,905,270

Table 7.4 Weighted normalized decision matrix

Criteria							
	P	IR	IP	OH	UL	O and M	tCO_2/y
A_1	0.018	0.012	0.005	0.014	0.037	0.005	0.038
A_2	0.035	0.012	0.005	0.014	0.037	0.005	0.063
A_3	0.088	0.012	0.005	0.014	0.037	0.005	0.190
A_4	0.018	0.020	0.008	0.019	0.046	0.005	0.009
A_5	0.071	0.009	0.010	0.012	0.046	0.002	0.005
A_6	0.124	0.008	0.013	0.012	0.046	0.002	0.005
A_7	0.177	0.066	0.010	0.016	0.046	0.014	0.010
A_8	0.018	0.024	0.005	0.045	0.027	0.023	0.050
A_9	0.018	0.024	0.005	0.045	0.027	0.017	0.050
A_{10}	0.018	0.024	0.005	0.045	0.027	0.017	0.050
A_{11}	0.018	0.024	0.005	0.045	0.027	0.009	0.050
A_{12}	0.198	0.011	0.005	0.045	0.037	0.015	0.095
A_{13}	0.007	0.020	0.008	0.042	0.037	0.008	0.116

Eq. 7.7, the relative closeness of every alternative to the ideal solution is calculated. The separation measures and the relative closeness to the ideal solution are shown in Table 7.6.

Finally, the ranking of the alternatives sorting by the value C_j^* is performed. The best alternative is one which has the shortest distance to the ideal solution and longest distance to the ideal solution. The results shown in Table 7.7 indicate that the Biomass plant alternative (Co-combustion in a conventional power plant, $P \geq 50Mw$) is the best choice, followed by the Windpower ($10 \leq P \leq 50Mw$) and Solar Thermoelectric ($P \geq 10Mw$) alternatives.

Table 7.5 Ideal (v_i^*) and negative-ideal solutions (v_i^-)

Criteria							
	P	*IR*	*IP*	*OH*	*UL*	O and M	tCO_2/y
	Max	*Min*	*Min*	*Max*	*Max*	*Min*	*Max*
v_i^*	0.198	0.008	0.005	0.045	0.046	0.002	0.190
v_i^-	0.018	0.066	0.013	0.012	0.027	0.023	0.005

Table 7.6 Separation measures and relative closeness to the ideal solution

	A_1	A_2	A_3	A_4	A_5	A_6	A_7	A_8	A_9	A_{10}	A_{11}	A_{12}	A_{13}
D_j^*	0.249	0.221	0.135	0.271	0.248	0.232	0.218	0.242	0.241	0.241	0.241	0.120	0.223
D_j^-	0.066	0.084	0.206	0.054	0.082	0.124	0.161	0.070	0.070	0.070	0.071	0.212	0.126
C_j^*	0.211	0.276	0.605	0.165	0.249	0.349	0.424	0.225	0.226	0.226	0.229	0.639	0.361

Table 7.7 Ranking

	A_{12}	A_3	A_7	A_{13}	A_6	A_2	A_5	A_{11}	A_9	A_{10}	A_8	A_1	A_4
C_j^*	0.639	0.605	0.424	0.361	0.349	0.276	0.249	0.229	0.226	0.226	0.225	0.211	0.165

7.3 Concluding Remarks

The main drawback of the TOPSIS method is that it introduces the ranking index including the distances from the ideal point and from the negative-ideal point without considering their relative importance, an issue that will be solved by using the VIKOR method in the following chapter. Applications of the TOPSIS method in the literature are very limited. To quote a few examples, the method has been applied to solid waste management [2], fuel buses for public transportation [3], sustainability of renewable energy options [4], the assessment of offshore wind turbine support structures [5] and in uncertain environments, the method has been applied to solar systems and energy planning [6, 7].

References

1. Huang CL, Yoon K (1981) Multi attribute decision making: methods and applications. Springer, New York
2. Cheng S, Chan C, Huang GH (2003) An integrated multi-criteria decision analysis and inexact mixed integer linear programming approach for solid waste management. Eng Appl Artif Intel 16(5–6):543–554
3. Tzeng GH, Lin CW, Opricovic S (2005) Multi-criteria analysis of alternative-fuel buses for public transportation. Energ Policy 33(11):1373–1383

4. Doukas H, Karakosta C, Psarras J (2010) Computing with words to assess the sustainability of renewable energy options. Expert Syst Appl 37(7):5491–5497
5. Lozano-Minguez E, Kolios AJ, Brennan FP (2011). Multi-criteria assessment of offshore wind turbine support structures. Renew Energ doi:10.1016/j.renene.2011.04.020
6. Cavallaro F (2010) Fuzzy TOPSIS approach for assessing thermal-energy storage in concentrated solar power (CSP) systems. Appl Energ 87(2):496–503
7. Kaya T, Kahraman C (2011) Multicriteria decision making in energy planning using a modified fuzzy TOPSIS methodology. Expert Syst Appl 38(6):6577–6585

Chapter 8
VIKOR

The Compromise Ranking Method, also known as the VIKOR method (VlseKriterijumska Optimizacija I Kompromisno Resenje in serbian, means Multicriteria Optimization and Compromise Solution), is an effective tool in multi-criteria decision making. This method introduces the multi-criteria ranking index based on the particular measure of 'closeness' to the 'ideal' solution, F^*. The compromise solution Fc is a feasible solution that is the 'closest' to the ideal solution, and compromise means an agreement established by mutual concessions. In this chapter, the method is applied to the selection among 13 different renewable energy options for electric generation.

8.1 VIKOR

Assuming that each alternative is evaluated according to each criterion function, the compromise ranking could be performed by comparing the measure of closeness to the ideal alternative and a compromise means an agreement established by mutual concessions. The multi-criteria measure for compromise ranking is developed from the Lp-metric used as an aggregating function in a compromise programming method [1, 2]:

$$L_{p,j} = \{\sum_{i=1}^{n}[\frac{w_i\left(f_i^* - f_{ij}\right)}{\left(f_i^* - f_i^-\right)}]^p\}^{\frac{1}{p}} \tag{8.1}$$

where $L_{1,j}$ and $L_{\infty,j}$ are used to formulate ranking measure; f_{ij} is the value of the i_{th} criterion function for the alternative A_j; and n is the number of criteria. The compromise ranking algorithm VIKOR is implemented carrying the following steps [3]:

J. R. San Cristóbal Mateo, *Multi-Criteria Analysis in the Renewable Energy Industry*, 49
Green Energy and Technology, DOI: 10.1007/978-1-4471-2346-0_8,
© Springer-Verlag London Limited 2012

Step 1. Determine the best and the worst values of all criterion functions

The best (f_i^*) and the worst (f_i^-) values of all criterion functions are determined, taking into account that, if the i_{th} function represents a benefit then:

$$\begin{cases} f_i^* = max_j f_{ij} \\ f_i^- = min_j f_{ij} \end{cases} \tag{8.2}$$

while if the i_{th} function represents a cost then:

$$\begin{cases} f_i^* = min_j f_{ij} \\ f_i^- = max_j f_{ij} \end{cases} \tag{8.3}$$

Step 2. Compute the values S_j and R_j

The values S_j and R_j are calculated by the relations:

$$S_j = \sum_{i=1}^n \frac{w_i(f_i^* - f_{ij})}{(f_i^* - f_i^-)} \tag{8.4}$$

and

$$R_j = max_i \frac{w_i(f_i^* - f_{ij})}{(f_i^* - f_i^-)} \tag{8.5}$$

where w_i are the weights of criteria, expressing the decision-maker's preference as the relative importance of the criteria.

Step 3. Compute the values Q_j

The values Q_j are calculated by the relation

$$Q_j = v\frac{S_j - S^*}{S^- - S^*} + (1 - v)\frac{(R_j - R^*)}{(R^- - R^*)} \tag{8.6}$$

where $S^* = min_j S_j$; $S^- = max_j S_j$; $R^* = min_j R_j$; $R^- = max_j R_j$ and v is introduced as a weight for the strategy of maximum group utility, whereas $(1 - v)$ is the weight of the individual regret. The solution obtained by $min_j S_j$ is with a maximum group utility ('majority' rule), and the solution obtained by $min_j R_j$ is with a minimum individual regret of the 'opponent'.

Step 4. Rank the alternatives

Rank the alternatives, sorting by the values S, R and Q in decreasing order. The results are three ranking lists.

Step 5. Propose the compromise solution

Propose as a compromise solution the alternative $A^{(1)}$ which is the best ranked by the measure Q (minimum), if the following two conditions are satisfied:

Table 8.1 Best (f_i^*) and worst (f_i^-) values of all criteria

	Criteria						
	P	IR	IP	OH	UL	OM	tCO_2/y
	Max	Min	Min	Max	Max	Min	Max
f_i^*	56,000	601	1	7,500	25	7.11	9,649,680
f_i^-	2,000	5,000	2.5	2,000	15	0.60	255,490

1. Acceptable advantage. $Q(A^2) - Q(A^1) \geq DQ$, where $DQ = \frac{1}{(m-1)}$ and A^2 is the alternative with second position in the ranking list by Q.
2. Acceptable stability in decision-making. The alternative A^1 must also be the best ranked by S and/or R. This compromise solution is stable within a decision making process, which could be the strategy of maximum group utility (when $v > 0.5$ is needed), or 'by consensus' ($v \approx 0.5$).

If one of the conditions is not satisfied, then a set of compromise solutions is proposed which consists of:

- Alternative A^1 and A^2 if only condition 2 is not satisfied, or
- Alternatives A^1, A^2, A^M if the condition 1 is not satisfied. A^M is determined by the relation $Q(A) - A(A^1) < DQ$ for maximum M (the positions of these alternatives are in 'closeness').

8.2 Application

Let us consider as a multi-criteria problem, the same problem as in the TOPSIS method, the selection among 13 different RE alternatives for electric generation evaluated according to the same criteria (Table 7.1). Of the criteria considered, Power, Operating Hours, Useful Life and Tons of emissions avoided are beneficial attributes and so higher values are desirable. Investment Ratio, Implementation Period and Operating and Maintenance Costs are non-beneficial attributes and so lower values are desirable. Table 8.1 shows the the best f_i^* and the worst f_i^- values of all criteria.

The values of S_j, R_j and Q_j obtained using Eqs. 8.3, 8.4, and 8.5 respectively are shown in Table 8.2. Ranking the alternatives, sorting by the values S, R and Q in decreasing order (Table 8.3) gives us, as a compromise solution, the alternative A_{12}. This alternative, a Biomass plant (Co-combustion in a conventional power plant) of $P \geq 50$ Mw is the best ranked by the measure Q (minimum). In addition, conditions 1 and 2 are satisfied as this alternative is also the best ranked by S and R.

When comparing the results obtained with the TOPSIS and VIKOR methods, we can see that the same alternative, the Co-combustion in a conventional power plant, is the highest ranked by both methods. Being the highest by the TOPSIS method means that this alternative is the best in terms of the ranking index. In addition, being the highest ranked alternative by the VIKOR method means that it

Table 8.2 Values of S_j, R_j and Q_j

	A_1	A_2	A_3	A_4	A_5	A_6	A_7	A_8	A_9	A_{10}	A_{11}	A_{12}	A_{13}
S_j	0.713	0.646	0.371	0.693	0.621	0.539	0.536	0.709	0.698	0.698	0.681	0.238	0.545
R_j	0.301	0.272	0.183	0.301	0.273	0.273	0.266	0.301	0.301	0.301	0.301	0.140	0.319
Q_j	0.955	0.802	0.263	0.934	0.778	0.691	0.669	0.951	0.939	0.939	0.921	0	0.826

Table 8.3 Ranking

Q_j		R_j		S_j	
Q_{12}	0.000	R_{12}	0.093	S_{12}	0.131
Q_7	0.289	R_7	0.151	S_7	0.364
Q_6	0.330	R_6	0.175	S_6	0.373
Q_3	0.582	R_3	0.258	S_5	0.492
Q_5	0.611	R_5	0.300	S_3	0.526
Q_2	0.868	R_2	0.383	S_4	0.600
Q_4	0.882	R_1	0.425	S_{13}	0.606
Q_{13}	0.922	R_4	0.425	S_2	0.650
Q_{11}	0.934	R_8	0.425	S_{11}	0.659
Q_9	0.953	R_9	0.425	S_{10}	0.680
Q_{10}	0.953	R_{10}	0.425	S_9	0.680
Q_1	0.963	R_{11}	0.425	S_1	0.692
Q_8	0.965	R_{13}	0.450	S_8	0.694

is the closest to the ideal solution, and to be as close as possible to the ideal is the rationale of human choice.

8.3 Concluding Remarks

TOPSIS and VIKOR methods use different kinds of normalization to eliminate the units of criterion functions: the VIKOR method uses linear normalization whereas the TOPSIS method uses vector normalization. However, the main difference appears in the aggregating approaches. The VIKOR method introduces an aggregating function representing the distance from the ideal solution. This ranking index is an aggregation of all criteria, the relative importance of the criteria, and a balance between total and individual satisfaction. The TOPSIS method introduces the ranking index including the distances form the ideal point and from the negative-ideal point without considering their relative importance. Applications of the VIKOR method can be found in Sayadi et al. [4], Chang [5], Sanayei et al. [6], Bazzari et al. [7] and San Cristóbal [8].

References

1. Yu PL (1973) A class of solutions for group decision problems. Manage Sci 19(8):936–946
2. Zeleny M (1982) Multiple criteria decision making. McGraw-Hill, New York
3. Opricovic S, Tzeng GH (2004) Compromise solution by MCDM methods: a comparative analysis of VIKOR and TOPSIS. Eur J Oper Res 156:445–55
4. Sayadi MK, Heydari M, Shahanaghi K (2009) Extension of VIKOR method for decision making problem with interval numbers. Appl Math Model 33(5):2257–2262
5. Chang CL (2010) A modified VIKOR method for multiple criteria analysis. Environ Monit Assess 168(1–4):339–344
6. Sanayei A, Farid Mousavi S, Yazdankhah A (2010) Group decision making process for supplier selection with VIKOR under fuzzy environment. Expert Syst Appl 37(1):24–30
7. Bazzazi AA, Osanloo M, Karimi B (2011) Deriving preference order of open pit mines equipment through MADM methods: application of modified VIKOR method. Expert Syst Appl 38(3):2550–2556
8. San Cristóbal JR (2011) Multi-Criteria decision-making in the selection of a renewable energy project in Spain: The Vikor method. Renew Energ 36(2):498–502

Chapter 9
A Multi-Criteria Data Envelopment Analysis

To estimate the efficiency scores of companies, various methods have been developed during the past two decades. These methods are generally classified as parametric and non-parametric methods. In the parametric methods, a cost of production function is estimated, whereas in the non-parametric methods, it is not necessary to estimate the cost or production function. Stochastic Frontier Analysis (SFA) and Data Envelopment Analysis (DEA) are the major parametric and non-parametric models respectively. In this chapter a Multiple Criteria Data Envelopment Analysis (MCDEA) model is developed and applied to the selection of a renewable project based on the concept of efficiency.

9.1 DEA

Charnes et al. [1] first introduced the DEA concept and many studies have since appeared that deal with various types of applications where the presence of multiple inputs and outputs makes comparison difficult. A non-parametric piecewise frontier (a best practice efficiency frontier) composed of Decision-Making Units (*DMUs*), which own the optimal efficiency over the datasets is constructed by DEA for comparative efficiency measurement. Those *DMUs* located at the efficiency frontier have their maximum outputs generated among all *DMUs* by taking the minimum level of inputs, which are efficient *DMUs* and own the best efficiency among all *DMUs*. The existing gap from any *DMUs* to the efficiency frontier shows how far the *DMUs* should be further improved to reach the optimal efficiency level. DEA produces detailed information on the efficiency of the unit, to be measured without any assumptions regarding the functional form of the production function, not only relative to the efficiency frontier, but also to specific efficient units which can be identified as role models [2, 3]. Thus, DEA can be used by inefficient organizations to benchmark efficient and 'best-practice organizations'.

The classical DEA model for evaluating the efficiency of a *DMU*, denoted by DMU_0 is as follows [1]:

J. R. San Cristóbal Mateo, *Multi-Criteria Analysis in the Renewable Energy Industry*,
Green Energy and Technology, DOI: 10.1007/978-1-4471-2346-0_9,
© Springer-Verlag London Limited 2012

$$\max \ h_0 = \textstyle\sum_{r=1}^{s} u_r y_{rj_0}$$
$$s.t.$$
$$\sum_{i=1}^{m} v_i x_{ij_0} = 1 \tag{9.1}$$
$$\sum_{r=1}^{s} u_r y_{rj} - \sum_{i=1}^{m} v_i x_{ij_0} \le 0$$
$$u_r, v_i \ge 0$$

where j is the *DMU* index ($j = 1, ..., n$); r the output index ($r = 1, ..., s$); i the input index ($i = 1, ..., m$); x_{ij} the value of the ith input of the jth *DMU*; y_{rj} the value of the rth output of the jth *DMU*; u_r the weight given to the rth output; v_i the weight given to the ith input; and h_0 the relative efficiency of DMU_0, the *DMU* under evaluation. In this model, DMU_0 is efficient if and only if $h_0 = 1$.

Lii and Reeves [4] extend the concept of relative efficiency and the method of efficiency evaluation from single criterion-based conventional approach to multiple criteria-oriented one. The authors argue that the resulting DEA models are more flexible and powerful in many aspects, particularly in discriminant analysis and weight restriction, showing three advantages. First, in classical DEA, if a *DMU* is efficient, its optimal (weight) solution is almost surely non-unique. In this situation, a Linear Programming procedure returns the first optimal solution it finds usually giving solutions with extremely distributed weights. Second, Multi Objective Linear Programming (MOLP) solutions contain not only the solution that individually optimize each of the objectives, but also other non-dominated solutions that, while non-optimal under any given criterion, provides alternative choices for the associated *DMU*. There may be situations where some of these non-dominated solutions are more preferred to those solutions optimizing individual objectives. Third, the total number of non-dominated solutions associated with a *DMU* often reflects the stability of a *DMUs* efficiency scores relative to the changes in efficiency criteria.

Model (9.1) can be expressed equivalently in the following deviational variable form:

$$\min \ d_0 \left(\text{or} \max h_0 = \textstyle\sum_{r=1}^{s} u_r y_{rj_0} \right)$$
$$s.t.$$
$$\sum_{i=1}^{m} v_i x_{ij_0=1} \tag{9.2}$$
$$\sum_{r=1}^{s} u_r y_{rj} - \sum_{i=1}^{m} v_i x_{ij_0} + d_j = 0$$
$$u_r, v_i, d_j \ge 0$$

where d_0 is the deviation variable for DMU_0 and d_j the deviation variable for the jth *DMU* (appeared at the jth original inequality constraint). Under this model, DMU_0 is efficient if and only if $d_0 = 0$ (or equivalently, $h_0 = 1$). If DMU_0 is not efficient, its efficiency score is $h_0 = 1 - d_0$. The quantity d_0, which is bounded by the interval $[0, 1)$, can be regarded as a measure of inefficiency. The smaller the value of d_0, the less inefficient (and thus the more efficient) DMU_0 is. In this sense, we say that classical DEA method [Model (9.1) or (9.2)] minimizes DMU_0's inefficiency, as measured by d_0, under the constraint that the weighted sum of the outputs is less than or equal to the weighted sum of the inputs for each *DMU*.

Lii and Reeves [4] propose a MCDEA model that is different from other models in that each criterion is an independent objective function. That is, each criterion defines a particular concept of efficiency and there is no prior preference among these efficiency criteria. The form of the proposed MCDEA model depends upon the efficiency criteria used. For a MCDEA problem that has the following three criteria, i.e., minimizing d_0, minimizing the sum of the deviations and minimizing the maximum deviation, the MCDEA model can be specified as follows:

$$\min \, d_0 (\text{or max } h_0 = \sum_{r=1}^{s} u_r y_{rj_0})$$
$$\min \sum_{j=1}^{n} d_j$$
$$\min M$$
$$s.t.$$
$$\sum_{i=1}^{m} v_i x_{ij_0} = 1$$
$$\sum_{r=1}^{s} u_r y_{rj} - \sum_{i=1}^{m} v_i x_{ij_0} + d_j = 0$$
$$M - d_j \geq 0$$
$$u_r, v_i, d_j \geq 0$$

(9.3)

The first objective of model (9.3) is identical to the objective of model (9.1) or (9.2). The second objective function is a straightforward representation of the deviation sum and the variable M in the third objective represents the maximum quantity among all deviation variables d_j $(j = 1, ..., n)$. Notice that the feasible region for decision variables u_r and v_i in model (9.3) is the same as that in models (9.1) and (9.2). The effect of added constraints, $M - d_j \geq 0$ is to make M the maximum deviation. They do not change the feasible region of decision variables.

Model (9.3) is a Multi-objective linear programming model (MOLP) and, in a MOLP problem it is generally impossible to find a solution that optimizes all objectives simultaneously. Therefore, the task of a MOLP solution process is not to find an optimal solution but, instead, to find non-dominated solutions (in multiple criteria terminology, a non-dominated solution is also called an efficient solution) and to help select a most preferred one. One fact to point out is that a non-dominated solution set for a MOLP problem will always contain, but is not limited to the optimal solutions obtained by individually optimizing each of the objectives in the MOLP problem under the setting of single objective linear programming.

The solution that optimizes the first objective function of model (9.3) is equivalent to the optimal solution of model (9.1) or (9.2). That is, DMU_0 is efficient (in the classical sense) if and only if the value of d_0 corresponding to the solution that optimizes the first objective function of model (9.3) is zero. In the same way, it can be defined a DMU's relative efficiency corresponding to the second and third objective criteria in the following way: DMU_0 is minsum efficient and minimax efficient if and only if the value of d_0 corresponding to the solution that minimizes the second and third objective function of model (9.3) is zero. In all the three above definitions, no matter DMU_0 is efficient or not, its DEA efficiency score is $1 - d_0$.

Efficiencies defined under minsum and minimax criterion are most restrictive than that defined in the classical DEA and these two criteria generally yield fewer efficient DMUs. That is, it is more difficult for a DMU to achieve minsum or minimax efficiency that to achieve classical DEA efficiency. More precisely, if DMU_0 is minsum or minimax efficient, it must also be DEA efficient, because by definition, minsum or minimax efficiency requires $d_0 = 0$. However, if DMU_0 is DEA efficient, it may or may not be minsum or minimax efficient because d_0 does not necessarily imply that M or $\sum d_j$ is minimized. Including these new criteria in a DEA model will result in the improvement in discriminating power. On the other hand, since M or $\sum d_j$ are functions of all deviation variables and each deviation variable is related to a constraint, minimizing M or $\sum d_j$ is, in some sense, equivalent to improving tighter constraints on weight variables.

9.2 Application

In this section, the MCDEA model is applied to determine the most appropriate RE alternative based on the concept of efficiency [5]. The input and output data of 13 renewable energies technologies used to perform the analysis, shown in Table 9.1 are defined as follows:

Inputs	
v_1	Investment ratio ($Euro * 10^3/Kw$)
v_2	Implement period (years)
v_3	Operating and maintenance costs ($Euro * 10^{-3}/Kwh$)
Outputs	
u_1	Power ($Mw * 10^3$)
u_2	Operating hours (Hours $* 10^3$/year)
u_3	Useful life (years)
u_4	Tons of CO_2 avoided ($tCO_2 * 10^6$/year)

The results of classical DEA method, solved by LP are given in Table 9.2. Using model (9.3), the results obtained are shown in Tables 9.3, 9.4 and 9.5, corresponding to the three criteria respectively (minimizing d_0, minimizing the sum of the deviations and minimizing the maximum deviation). Efficiency scores in Tables 9.2 and 9.3 are identical since they are obtained under the same criterion. As we can see, all *DMUs* except DMU_7 (Solar Thermo-electric $P \geq 10 Mw$) are efficient. However, we can easily observe differences between the solutions shown in both Tables. The input and output weights obtained by the MOLP procedure in Table 9.3 are distributed more evenly than those obtained by classical DEA in Table 9.2.

Table 9.4 lists the efficiency scores for each *DMU* under the Minsum criteria. Under this criteria the only *DMUs* rated as efficient are DMU_1 (Windpower $P \leq 5 Mw$), DMU_2 (Windpower $5 \leq P \leq 10 Mw$), DMU_3 (Windpower $10 \leq P \leq 50 Mw$) and DMU_{11} (Biomass forest industrial wastes $P \leq Mw$). Table 9.5lists the efficiency scores for each *DMU* under the Minimax criterion. In particular, only DMU_3 (Windpower $10 \leq P \leq 50 Mw$) and DMU_{12} (Biomass Co-combustion in conventional central $P \geq 50 Mw$) are rated as efficient, while the rest of the *DMUs*, rated as efficient by classical DEA and by Minsum criteria, are no longer efficient

Table 9.1 Input and output data for the 13 *DMUs*

DMU		Input weights			Output weights			
		v_1	v_2	v_3	u_1	u_2	u_3	u_4
DMU_1	Windpower $P \leq 5\,Mw$	0.937	1	1.47	0.5	2.35	20	1.93
DMU_2	Windpower $5 \leq P \leq 10\,Mw$	0.937	1	1.47	1.0	2.35	20	3.22
DMU_3	Windpower $10 \leq P \leq 50\,Mw$	1.500	1	1.51	2.5	2.35	20	9.65
DMU_4	Hydroelectric $P \leq 10\,Mw$	0.700	1.5	1.45	0.5	3.10	25	0.47
DMU_5	Hydroelectric $10 \leq P \leq 25\,Mw$	0.601	2	0.70	2.0	2.00	25	0.26
DMU_6	Hydroelectric $25 \leq P \leq 50\,Mw$	5.000	2.5	0.60	3.5	2.00	25	0.26
DMU_7	Solar Thermo-electric $P \geq 10\,Mw$	1.803	2	4.2	5.0	2.59	25	0.48
DMU_8	Biomass (energetic cultivations) $P \leq 5\,Mw$	1.803	1	7.11	0.5	7.50	15	2.52
DMU_9	Biomass (forest and agricultural wastes) $P \leq 5\,Mw$	1.803	1	5.42	0.5	7.50	15	2.52
DMU_{10}	Biomass (farming industrial wastes) $P \leq 5\,Mw$	1.803	1	5.42	0.5	7.50	15	2.52
DMU_{11}	Biomass (forest industrial wastes) $P \leq 5\,Mw$	1.803	1	2.81	0.5	7.50	15	2.52
DMU_{12}	Biomass (Co-combustion in conventional central) $P \geq 50\,Mw$	0.856	1	4.56	5.6	7.50	20	4.84
DMU_{13}	Bio fuels $P \leq 2\,Mw$	1.503	1.5	2.51	0.2	7.00	20	5.91

Table 9.2 Classical DEA results

DMU	Efficiency	Input weights			Output weights			
		v_1	v_2	v_3	u_1	u_2	u_3	u_4
DMU_1	1	0.14	0.40	0.32	0	1.31	3.46	0
DMU_2	1	0.29	0.38	0.23	0	1.30	3.44	0
DMU_3	1	0	0.44	0.37	0.38	0	0	0
DMU_4	1	0.10	0.32	0.26	0.01	1.05	2.70	0
DMU_5	1	0.45	0.31	0.08	0.07	0	3.45	0
DMU_6	1	0	0.33	0.28	0.28	0	0	0
DMU_7	0.78	0	0.20	0.15	0.14	0	0.27	0
DMU_8	1	0	1.00	0	0	1.33	0	0
DMU_9	1	0	1.00	0	0	1.33	0	0
DMU_{10}	1	0	1.00	0	0	1.33	0	0
DMU_{11}	1	0	0.12	0.31	0.07	1.24	0	0.01
DMU_{12}	1	0	0.20	0.17	0.18	0	0	0
DMU_{13}	1	0.08	0.12	0.28	0.07	1.30	0	0.01

with respect to the Minimax criteria. As we can see, the only *DMU* rated as efficient under the three criteria is DMU_3 (Windpower $10 \leq P \leq 50\,Mw$) and it can be considered the only non-dominated solution.

Table 9.3 Minimizing d_0 results

DMU	Efficiency	Input weights			Output weights			
		v_1	v_2	v_3	u_1	u_2	u_3	u_4
DMU_1	1	0.10	9.02	0	0	0	5	0
DMU_2	1	0	7.89	0.14	0.06	0.54	3.64	0
DMU_3	1	0	7.84	0.14	0.06	0.54	3.62	0
DMU_4	1	0	5.67	0.10	0.04	0.39	2.62	0
DMU_5	1	0	4.70	0.08	0.04	0.32	2.17	0
DMU_6	1	0	3.83	0.07	0.03	0.26	1.77	0
DMU_7	0.78	0	3.62	0.07	0.03	0.25	1.67	0
DMU_8	1	0	3.01	0.10	0.02	0.48	1.37	0
DMU_9	1	0	3.61	0.12	0.03	0.57	1.65	0
DMU_{10}	1	0	3.61	0.12	0.03	0.57	1.65	0
DMU_{11}	1	0	6.15	0.12	0.05	0.45	3.05	0
DMU_{12}	1	0	4.02	0.13	0.03	0.64	1.83	0
DMU_{13}	1	0	5.11	0.09	0.04	0.35	2.36	0

Table 9.4 Minsum DEA results

DMU	Efficiency	Input weights			Output weights			
		v_1	v_2	v_3	u_1	u_2	u_3	u_4
DMU_1	1	0	8.51	0.10	0	0.67	4.21	0
DMU_2	1	0	8.51	0.10	0	0.67	4.21	0
DMU_3	1	0	8.48	0.10	0	0.67	4.19	0
DMU_4	0.88	0	5.98	0.07	0	0.47	2.96	0
DMU_5	0.67	0	4.80	0.06	0	0.38	2.37	0
DMU_6	0.54	0	3.89	0.05	0	0.31	1.92	0
DMU_7	0.36	0.10	1.10	0.06	0	0.38	1.04	0
DMU_8	0.53	0.06	1.10	0.11	0.02	0.54	0.74	0
DMU_9	0.65	0.08	1.34	0.13	0.03	0.70	0.91	0
DMU_{10}	0.65	0	3.61	0.12	0.03	0.57	1.65	0
DMU_{11}	1	0.15	3.26	0.14	0	0.86	2.36	0
DMU_{12}	0.80	0	1.80	0.18	0.02	1.64	1.28	0
DMU_{13}	0.86	0	5.56	0.07	0	0.44	2.75	0

9.3   Concluding Remarks

Application of DEA in the electric power industry may be traced back to the publications of Cote [6], Hjalmarsson and Veiderpass [7], Miliotis [8] and Golany et al. [9]. Since then, DEA has been widely applied both to the distribution aspects of operations and to the electric generation. Jha and Shrestha [10] evaluated the efficiency of Hydropower plants in Nepal using DEA and Vaninski [11] using DEA estimated the efficiency of electric power generation in the U.S. for the period of 1991 through 2004. Azadeh et al. [12] presented and integrated hierarchical DEA-Principal

Table 9.5 Minimax DEA results

DMU	Efficiency	Input weights			Output weights			
		v_1	v_2	v_3	u_1	u_2	u_3	u_4
DMU_1	0.88	0.10	9.02	0	0	0	5	0
DMU_2	0.91	0	7.89	0.14	0.06	0.54	3.64	0
DMU_3	1	0	7.84	0.14	0.06	0.54	3.62	0
DMU_4	0.79	0	5.67	0.10	0.04	0.39	2.62	0
DMU_5	0.68	0	4.70	0.08	0.04	0.32	2.17	0
DMU_6	0.60	0	3.83	0.07	0.03	0.26	1.77	0
DMU_7	0.62	0	3.62	0.07	0.03	0.25	1.67	0
DMU_8	0.58	0	3.01	0.10	0.02	0.48	1.37	0
DMU_9	0.70	0	3.61	0.12	0.03	0.57	1.65	0
DMU_{10}	0.70	0	3.61	0.12	0.03	0.57	1.65	0
DMU_{11}	0.82	0	6.15	0.12	0.05	0.45	3.05	0
DMU_{12}	1	0	4.02	0.13	0.03	0.64	1.83	0
DMU_{13}	0.72	0	5.11	0.09	0.04	0.35	2.36	0

Component Analysis approach for location of solar plants. Sadjadi and Omrani [13] presented a DEA model with uncertain data for performance assessment of Iranian electricity distribution companies. Jayanthi et al. [14] showed a DEA-based application to the U.S. photovoltaic industry to evaluate the potential of innovations occurring in various stages of the industry value chain in terms of its relative efficiency with respect to a best practices frontier.

References

1. Charnes A, Cooper WW, Rhodes E (1978) Measuring the efficiency of decision-making units. Eur J Oper Res 2:429–444
2. Hu JL, Kao CH (2007) Efficient energy-savings targets for APEC economies. Energ Policy 35:373–382
3. Hu JL, Wang SC, Yeh FY (2006) Total-factor water efficiency of regions in China. Res Policy 31:217–230
4. Li XB, Reeves GR (1999) A multiple criteria approach to data envelopment analysis. Theory and Methodology. Eur J Oper Res 115:507–517
5. San Cristóbal JR (2011) A multi-criteria data envelopment analysis model to evaluate the efficiency of the renewable energy technologies. Renew Energ 36(10):2742–2746
6. Cote DO (1989) Firm efficiency and ownership structure-the case of U.S. electric utilities using panel data. Ann Public Coop Econ 60(4):431–450
7. Hjalmarsson L, Veiderpass A (1992) Efficiency and ownershipin Swedish electricity retail distribution. J Prod Anal 3(1/2):7–23
8. Milliotis PA (1992) Data envelopment analysis applied to electricity distribution districts. J Oper Res Soc 43:549–555
9. Golany B, Roll Y, Ryback D (1994) Measuring efficiency of power plants in Israel by data envelopment analysis. IEEE T Eng Manage 41(3):291–301

10. Jha DK, Shrestha R (2006) Measuring efficiency of hydropower plants in Nepal using data envelopment analysis. IEEE T Power Syst 21(4):1502–1511
11. Vaninsky A (2006) Efficiency of electric power generation in the United States: analysis and forecast based on data envelopment analysis. Energ Econ 28(3):326–338
12. Azadeh A, Ghaderi SF, Maghsoudi A (2008) Location optimization of solar plants by an integrated hierarchical DEA PCA approach. Energ Policy 36:3393–4004
13. Sadjadi SJ, Omrani H (2008) Data envelopment analysis with uncertain data: a application for Iranian electricity distribution companies. Energ Policy 36:4247–4254
14. Jayanthi S, Witt EC, Singh V (2009) Evaluation of potential innovations: a DEA-based application to U.S. photovoltaic industry. IEEE T Eng Manage 56(3):478 August

Chapter 10
Multi-Attribute Utility Theory

Multi-attribute utility theory is an extension of Utility theory developed to help decision-makers assign utility values, taking into consideration the decision-maker's preferences, to outcomes by evaluating these in terms of multiple attributes and combining these individual assignments to obtain overall utility measures. Utility theory has generally been used to develop a relationship between utility and costs incurred as a consequence of a particular decision. There are situations where, rational decision-makers who subscribe the von Neuman-Morgenstern axioms are sometimes willing to violate the Expected Monetary Value minimization criterion (when dealing with benefits to maximize the Expected Monetary Value) and to sacrifice it to reduce risk, choosing the alternative that maximizes his or her expected utility.

10.1 The MAU Function

Multi-criteria utility theory generally combines the main advantages of simple scoring techniques and optimization models. Further, in situations in which satisfaction is uncertain, utility functions have the property that expected utility can be used as a guide to rational decision-making.

Multi-attribute utility theory takes into consideration the decision-maker's preferences in the form of utility function which is defined over a set of attributes. The utility value can be determined in the following three steps [1]: (i) Determination of single attribute utility functions. The values of utilities can vary between zero and one and reflect the level of importance in the achievement of that attribute; (ii) Verification of preferential and utility independence conditions and (iii) Derivation of the multi-attribute utility function.

A utility function is a device which quantifies the preferences of a decision-maker by assigning a numerical index to varying levels of satisfaction of a particular criterion. For a single criterion, the utility of satisfaction of a consequence x' is denoted by $(u(x'))$. Utility functions are constructed such that $(u(x'))$ is less preferred to $(u(x''))$, i. e.

J. R. San Cristóbal Mateo, *Multi-Criteria Analysis in the Renewable Energy Industry*, Green Energy and Technology, DOI: 10.1007/978-1-4471-2346-0_10, © Springer-Verlag London Limited 2012

Fig. 10.1 Utility functions

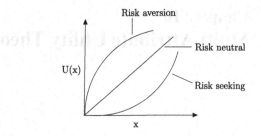

$$u(x') < u(x'')$$

if and only if x' is less preferred to x'', i. e. $x' < x''$

In other words, a utility function is a transformation of some level of project (alternative) performance, x', measured in its natural units into an equivalent level of decision-maker satisfaction. Utility functions contain information about the decision-maker's attitude toward risk being reflected in the shape of the utility curve which combines the decision-maker's preference attitudes, i.e. increasing or decreasing utility with increasing or decreasing x'. Depending on the decision maker's attitudes toward risk, utility functions can be concave, convex or linear as shown in Fig. 10.1. Concave utility functions describe risk-averse situations, convex utility functions describe risk-seeking situations and linear utility functions describe a risk-neutral situation.

All decisions involve choosing one, from several alternatives. Typically, each alternative is assessed for desirability on a number of scored criteria. What connects the criteria scores with desirability is the utility function. The most common formulation of a utility function is the additive model:

$$U_i = W_j, U_{ij} \text{ for all } i \tag{10.1}$$

where U_i is the overall utility value of alternative i; U_{ij} is the utility value of the jth criterion for the ith alternative and U_{ij} equals $u(X_i)$, for $1 \geq i \geq n$ and $1 \geq j \geq m$. X_i designates a specific value of x_{ij}; n is the total number of criteria, m is the total number of alternatives and W_j is the relative weight of the jth criterion.

The advantage of the additive form is its simplicity. In order to determine the overall utility function for any alternative, a decision-maker needs to determine only the n utility functions for that alternative. For the construction of the utility functions the decision-maker's preferences are analyzed by the method suggested by Bell et al. [2], and Keeney and Raiffa [3]. The first step involves the identification of the best and the worst outcomes (criteria scores) for each one of the criteria. The decision-maker is free to set these utility values at any level provided that the best outcome has the higher value. The usual method is to assign the worst outcome a utility value of zero and the best outcome a utility value of unity. This establishes the range of utility values to from 0 to 1 between the worst and the best possible outcomes. To determine the utility of intermediate values, the decision-maker is offered the following options:

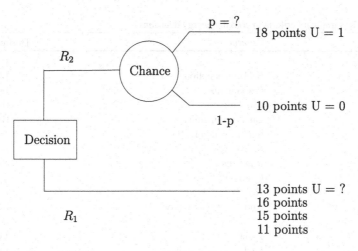

Fig. 10.2 Routes

1. Certain option. In this case the decision-maker is offered a certain outcome with a probability $p = 1$;
2. Risk option. In this case the decision-maker is offered a probabilistic outcome in the form of a gamble, in which the decision-maker either receives the best outcome with a probability p or the worst outcome with a probability of $1 - p$.

The following is an example of how the utility values for an hypothetical criterion are obtained and from which the utility curve is established. Let us consider the following scores of five alternatives for an hypothetical criterion:

	A	B	C	D	E
Scores	13	14	10	18	16

The first step is to identify the best and worst outcomes for the criterion considered and assign arbitrary utility values of 1 for the best outcome (alternative D with 18 points) and 0 for the worst outcome (alternative C with 10 points). The utility of the intermediate values is then determined by offering the decision-maker a choice between the following lotteries (Fig. 10.2):

1. Lottery 1: go to route R_1 for a certain consequence of 13 points (alternative A).
2. Lottery 2: go to route R_2 for either a best consequence of 18 points (alternative D) with a probability of p or a worst consequence of 10 points (alternative C) with a probability of $1 - p$.

What utility value should the decision-maker assign to a certain outcome of 13 points? For the decision-maker to make good decision and choose from the two routes, the utility value of 13-point score must be assessed and compared with the

Table 10.1 Significant criteria in cooking energy decisions

Number	Criteria	Desired value
Technical		
C_1	Fuel consumption	Low
C_2	Cooking time	Low
C_3	Durability	High
C_4	Quality, reliability	High
C_5	Sophistication level	High
C_6	Size/Weight/Space needs	Low
C_7	Ruggedness	High
C_8	Continuity of use	High
C_9	Need for tracking	Low
C_{10}	Nutrition value of food	High
Economic		
C_{11}	Initial cost	Low
C_{12}	Fuel cost per month	Low
C_{13}	Maintenance cost	Low
C_{14}	Available subsidy	Low
C_{15}	Rate of interest on loan, if any	Low
Social		
C_{16}	Pollution hazards	Low
C_{17}	Human drudgery	Low
C_{18}	Overall safety	High
Behavioral		
C_{19}	Aesthetics	High
C_{20}	Motivation to buy	High
C_{21}	Taste of food	High
C_{22}	Cleanliness of utensils	High
C_{23}	Ease of operation	High
C_{24}	Type of dishes cooked	High
C_{25}	Need for additional cooking system	Low
Commercial		
C_{26}	Improvements in models	High
C_{27}	Spares and after sales services	High
C_{28}	Distribution network	High
C_{29}	Market research	High
C_{30}	Need for user training	Low

expected utility of the risk option. To do this, the decision-maker determines a relative preference for a 13-point consequence by finding the probability p for the best outcome, to which the decision-maker is indifferent, between the certain route R_1 for a 13-point outcome and the gamble route R_2 for the two possible outcomes of 18 and 10 points. Let us assume that there is a probability of 0.3 for getting the best outcome and a probability of 0.7 for getting the worst outcome from the route R_2. Which route would the decision-maker prefer in this case? Since $p = 0.3$ the chance of getting the best outcome from route R_2 is very small, so in this case the decision-maker will

Table 10.2 Alternatives

Alternative	Description
A_1	Chulha
A_2	Improved chulha
A_3	Kerosene stove
A_4	Biogas stove
A_5	LPG stove
A_6	Microwave oven
A_7	Electric oven
A_8	Solar box cooker
A_9	Parabolic solar cooker

not gamble. He prefers to choose route R_1 with a 13-point certain outcome. Now, let us assume that there is a probability of 0.9 for getting the best outcome and a probability 0.1 for getting the worst outcome from route R_2. Which route does the decision-maker now prefer?. Since $p=0.9$, in this case there is a high chance of getting the best outcome of 18 points, so he will gamble and choose route R_2. Now, let us take a probability of 0.45 for getting the best outcome and a probability of 0.55 for getting the worst outcome from route R_2. Which route does he or she now prefer? Putting $p=0.45$ makes the thing difficult to choose for the decision-maker, but he will go for the certain outcome route R_1. Doing some more of these trials and errors, the decision-maker considers that a probability of 0.5 will make him indifferent between the two routes R_1 and R_2. According to utility theory, by choosing the probability that makes him indifferent between the two routes, the decision-maker has assigned a utility value for the certain outcome of 13 points. It is known from the principles of probabilities that the expected value of any random variable in the space will equal the sum of probability of each variable times its score. In this case, the expected utility for the route R_2 which includes two variables or two outcomes (the best outcome with $u = 1$ and the worst outcome with $u = 0$) will be:

$$p(\text{utility of best outcome}) + (1 - p)(\text{utility of worst outcome})$$
$$= 0.5 * U(18) + (1 - 0.5)U(10) = 0.5 * 1 + 0.5 * 0$$

Since the decision-maker is indifferent between 13 points for certain and this gamble, the alternatives must have the same utility value that is, $U(13) = 0.5$. This procedure can be used for any scores between 10 and 18. The more utility values obtained, the better the utility curves appear.

10.2 Application

Let us consider the case by Pohekar and Ramachandran [1] of a parabolic solar cooker (PSC) which is evaluated with respect to eight prevalent domestic cooking devices in India. Thirty different criteria categorized under technical, economic, environmental,

Table 10.3 Evaluation matrix for all criteria

Criteria	A_1	A_2	A_3	A_4	A_5	A_6	A_7	A_8	A_9	Scaling constant
C_1	2	1	0.5	0	0.25	2	2	0	0	0.0355
C_2	60	60	30	15	15	5	30	180	20	0.0384
C_3	1	4	15	5	20	5	5	10	20	0.0397
C_4	2	2	8	4	10	10	6	6	6	0.0376
C_5	2	2	6	6	10	10	10	4	4	0.0326
C_6	2	1	2	50	10	5	3	5	15	0.0332
C_7	2	2	6	6	10	8	8	8	8	0.0310
C_8	10	10	10	4	10	10	10	4	2	0.0335
C_9	0	0	0	0	0	0	0	4	10	0.0326
C_{10}	6	6	6	6	6	2	2	10	8	0.0371
C_{11}	10	50	200	5000	4000	8000	5000	2000	7000	0.0371
C_{12}	20	10	100	0	250	200	400	0	0	0.0293
C_{13}	0	0	50	200	50	200	200	50	20	0.0274
C_{14}	0	50	0	2000	0	0	0	500	2000	0.0252
C_{15}	0	0	0	0	13	13	13	3	3	0.0230
C_{16}	10	10	8	2	4	10	10	0	0	0.0217
C_{17}	10	10	6	6	2	2	2	2	2	0.0296
C_{18}	2	2	4	10	6	2	2	10	8	0.0433
C_{19}	2	2	4	4	10	10	8	8	8	0.0348
C_{20}	2	2	6	6	10	8	4	4	4	0.0314
C_{21}	6	6	2	6	6	2	10	10	10	0.0381
C_{22}	2	2	2	8	10	10	10	10	10	0.0368
C_{23}	10	10	10	4	8	4	4	6	4	0.0387
C_{24}	10	10	10	10	10	2	2	2	4	0.0332
C_{25}	2	2	4	4	6	8	10	10	10	0.0306
C_{26}	2	2	6	4	10	10	6	6	6	0.0342
C_{27}	10	10	10	4	10	6	4	4	2	0.0352
C_{28}	10	6	6	6	2	6	2	2	2	0.0355
C_{29}	2	2	8	2	10	10	4	4	4	0.0315
C_{30}	2	2	4	8	4	10	6	6	8	0.0323

social behavioral and commercial aspects are considered for the evaluation. To illustrate the use of different utility functions, the model presented by the authors, which only uses linear utility functions, is extended to concave and convex utility functions showing decision-maker's risk-aversion or risk-seeking attitude toward risk. Table 10.1 presents a summary of 30 criteria (twenty one criteria of qualitative nature and nine of qualitative nature) and their significance in evaluation of a cooking device. Table 10.2 shows the alternatives to evaluate and Table 10.3 the evaluation matrix for all criteria. The values of 2, 4, 6, 8 and 10 for qualitative criteria are allotted to indicate values of performance of a device as very low, low, moderate, high and very high, respectively. Scaling constants indicate weighted average importance of criteria.

Table 10.4 Normalized evaluation matrix linear utility function, risk neutral

Criteria	A_1	A_2	A_3	A_4	A_5	A_6	A_7	A_8	A_9
C_1	0.000	0.500	0.750	1.000	0.850	0.000	0.000	1.000	1.000
C_2	0.686	0.686	0.857	0.0943	0.943	1.000	0.857	0.000	0.914
C_3	0.000	0.158	0.737	0.210	1.000	0.210	0.210	0.474	1.000
C_4	0.000	0.000	0.750	0.250	1.000	1.000	0.500	0.500	0.500
C_5	0.000	0.000	0.500	0.500	1.000	1.000	1.000	0.250	0.250
C_6	0.980	1.000	0.980	0.000	0.816	0.918	0.960	0.918	0.714
C_7	0.000	0.000	0.500	0.500	1.000	0.750	0.750	0.750	0.750
C_8	1.000	1.000	1.000	0.250	1.000	1.000	1.000	0.250	0.000
C_9	1.000	1.000	1.000	1.000	1.000	1.000	1.000	0.600	0.000
C_{10}	0.500	0.500	0.500	0.500	0.500	0.000	0.000	1.000	0.750
C_{11}	1.000	0.995	0.976	0.375	0.501	0.000	0.375	0.751	0.125
C_{12}	0.950	0.975	0.750	1.000	0.375	0.500	0.000	1.000	1.000
C_{13}	1.000	1.000	0.750	0.000	0.750	0.000	0.000	0.750	0.900
C_{14}	1.000	0.980	1.000	1.000	0.000	0.000	0.000	0.750	1.000
C_{15}	1.000	1.000	1.000	0.769	0.000	0.000	0.000	0.769	0.769
C_{16}	0.000	0.000	0.200	0.800	0.600	0.000	0.000	1.000	1.000
C_{17}	0.000	0.000	0.500	0.500	1.000	1.000	1.000	1.000	1.000
C_{18}	0.000	0.000	0.250	1.000	0.500	0.000	0.000	1.000	0.750
C_{19}	0.000	0.000	0.250	0.250	1.000	1.000	0.750	0.750	0.750
C_{20}	0.000	0.000	0.500	0.500	1.000	0.750	0.250	0.250	0.250
C_{21}	0.500	0.500	0.000	0.500	0.500	0.000	1.000	1.000	1.000
C_{22}	0.000	0.000	0.000	0.750	1.000	1.000	1.000	1.000	1.000
C_{23}	1.000	1.000	1.000	0.000	0.750	0.000	0.000	0.330	0.000
C_{24}	1.000	1.000	1.000	1.000	1.000	0.000	0.000	0.000	0.250
C_{25}	1.000	1.000	0.750	0.750	0.500	0.250	0.000	0.000	0.000
C_{26}	0.000	0.000	0.500	0.250	1.000	1.000	0.500	0.500	0.500
C_{27}	1.000	1.000	1.000	0.250	1.000	0.500	0.250	0.250	0.000
C_{28}	1.000	0.500	0.500	0.500	0.000	0.500	0.000	0.000	0.000
C_{29}	0.000	0.000	0.750	0.000	1.000	1.000	0.250	0.250	0.250
C_{30}	1.000	1.000	0.750	0.250	0.750	0.000	0.500	0.500	0.750

Table 10.4 shows the normalized evaluation matrix and utility for all the criteria and the alternatives considered using linear utility functions. The best outcome of each one of the criteria has been assigned a utility value of 1 and the worst outcome a utility value of zero. Since there are two kinds of criteria; the maximization criteria (the maximum value is desirable) and the minimization criteria (the minimum value is desirable) the intermediate utility values are obtained by normalizing the evaluation matrix as follows:

1. For maximization criteria

$$u_j = \frac{A_j - A_{\min}}{A_{\max} - A_{\min}} \tag{10.2}$$

Table 10.5 Normalized evaluation matrix concave utility function, risk aversion

Criteria	A_1	A_2	A_3	A_4	A_5	A_6	A_7	A_8	A_9
C_1	0.249	0.944	1.000	0.929	0.999	0.249	0.249	0.929	0.929
C_2	1.000	1.000	1.000	0.967	0.967	0.929	1.000	0.249	0.982
C_3	0.249	0.545	1.000	0.628	0.929	0.628	0.628	0.925	0.929
C_4	0.249	0.249	1.000	0.685	0.929	0.929	0.944	0.944	0.944
C_5	0.249	0.249	0.944	0.944	0.929	0.929	0.929	0.685	0.685
C_6	0.943	0.929	0.944	0.249	1.000	0.980	0.957	0.980	1.000
C_7	0.249	0.249	0.944	0.944	0.929	1.000	1.000	1.000	1.000
C_8	0.929	0.929	0.929	0.685	0.929	0.929	0.929	0.685	0.249
C_9	0.929	0.929	0.929	0.929	0.929	0.929	0.929	0.998	0.249
C_{10}	0.944	0.944	0.944	0.944	0.944	0.249	0.249	0.929	1.000
C_{11}	0.929	0.933	0.946	0.837	0.944	0.249	0.837	1.000	0.490
C_{12}	0.962	0.947	1.000	0.929	0.837	0.944	0.249	0.929	0.929
C_{13}	0.929	0.929	1.000	0.249	1.000	0.249	0.249	1.000	0.989
C_{14}	0.929	0.943	0.929	0.929	0.249	0.249	0.249	1.000	0.929
C_{15}	0.929	0.929	0.929	1.000	0.249	0.249	0.249	1.000	1.000
C_{16}	0.249	0.249	0.612	1.000	0.998	0.249	0.249	0.929	0.929
C_{17}	0.249	0.249	0.944	0.944	0.929	0.929	0.929	0.929	0.929
C_{18}	0.249	0.249	0.685	0.929	0.944	0.249	0.249	0.929	1.000
C_{19}	0.249	0.249	0.685	0.685	0.929	0.929	1.000	1.000	1.000
C_{20}	0.249	0.249	0.944	0.944	0.929	1.000	0.685	0.685	0.685
C_{21}	0.944	0.944	0.249	0.944	0.944	0.249	0.929	0.929	0.929
C_{22}	0.249	0.249	0.249	1.000	0.929	0.929	0.929	0.929	0.929
C_{23}	0.929	0.929	0.929	0.249	1.000	0.249	0.249	0.787	0.249
C_{24}	0.929	0.929	0.929	0.929	0.929	0.249	0.249	0.249	0.685
C_{25}	0.929	0.929	1.000	1.000	0.944	0.685	0.249	0.249	0.249
C_{26}	0.249	0.249	0.944	0.685	0.929	0.929	0.944	0.944	0.944
C_{27}	0.929	0.929	0.929	0.685	0.929	0.944	0.685	0.685	0.249
C_{28}	0.929	0.944	0.944	0.944	0.249	0.944	0.249	0.249	0.249
C_{29}	0.249	0.249	1.000	0.249	0.929	0.929	0.685	0.685	0.685
C_{30}	0.929	0.929	1.000	0.685	1.000	0.249	0.944	0.944	1.000

1. For minimization criteria

$$u_j = \frac{A_{\max} - A_j}{A_{\max} - A_{\min}} \tag{10.3}$$

where A_j represents the score assigned to cooking device in evaluation matrix. $A_{\max}$ and $A_{\min}$ are the maximum and minimum scores assigned for the selected criteria for the identified devices.

Tables 10.5 and 10.6 show utility values using concave (exhibiting risk-averse behavior) and convex utility functions (exhibiting risk-seeking behavior), respectively. The final utilities of the devices are computed by multiplying the normalized utility values by respective scaling constants and the final utility value is found by

Table 10.6 Normalized evaluation matrix convex utility function, risk seeking

Criteria	A_1	A_2	A_3	A_4	A_5	A_6	A_7	A_8	A_9
C_1	0.050	0.138	0.407	0.827	0.598	0.050	0.050	0.827	0.827
C_2	0.324	0.324	0.569	0.718	0.718	0.827	0.569	0.050	0.666
C_3	0.050	0.013	0.389	0.014	0.827	0.014	0.014	0.118	0.827
C_4	0.050	0.050	0.407	0.019	0.827	0.827	0.138	0.138	0.138
C_5	0.050	0.050	0.138	0.138	0.827	0.827	0.827	0.019	0.019
C_6	0.787	0.827	0.787	0.050	0.504	0.674	0.748	0.674	0.360
C_7	0.050	0.050	0.138	0.138	0.827	0.407	0.407	0.407	0.407
C_8	0.827	0.827	0.827	0.019	0.827	0.827	0.827	0.019	0.050
C_9	0.827	0.827	0.827	0.827	0.827	0.827	0.827	0.228	0.050
C_{10}	0.138	0.138	0.138	0.138	0.138	0.050	0.050	0.827	0.407
C_{11}	0.827	0.817	0.781	0.060	0.138	0.050	0.060	0.409	0.015
C_{12}	0.731	0.779	0.407	0.827	0.059	0.138	0.050	0.827	0.827
C_{13}	0.827	0.827	0.407	0.050	0.407	0.050	0.050	0.407	0.641
C_{14}	0.827	0.788	0.827	0.827	0.050	0.050	0.050	0.407	0.827
C_{15}	0.827	0.827	0.827	0.434	0.050	0.050	0.050	0.434	0.434
C_{16}	0.050	0.050	0.013	0.479	0.228	0.050	0.050	0.827	0.827
C_{17}	0.050	0.050	0.138	0.138	0.827	0.827	0.827	0.827	0.827
C_{18}	0.050	0.050	0.019	0.827	0.138	0.050	0.050	0.827	0.407
C_{19}	0.050	0.050	0.019	0.019	0.827	0.827	0.407	0.407	0.407
C_{20}	0.050	0.050	0.138	0.138	0.827	0.407	0.019	0.019	0.019
C_{21}	0.138	0.138	0.050	0.138	0.138	0.050	0.827	0.827	0.827
C_{22}	0.050	0.050	0.050	0.407	0.827	0.827	0.827	0.827	0.827
C_{23}	0.827	0.827	0.827	0.050	0.407	0.050	0.050	0.040	0.050
C_{24}	0.827	0.827	0.827	0.827	0.827	0.050	0.050	0.050	0.019
C_{25}	0.827	0.827	0.407	0.407	0.138	0.019	0.050	0.050	0.050
C_{26}	0.050	0.050	0.138	0.019	0.827	0.827	0.138	0.138	0.138
C_{27}	0.827	0.827	0.827	0.019	0.827	0.138	0.019	0.019	0.050
C_{28}	0.827	0.138	0.138	0.138	0.050	0.138	0.050	0.050	0.050
C_{29}	0.050	0.050	0.407	0.050	0.827	0.827	0.019	0.019	0.019
C_{30}	0.827	0.827	0.407	0.019	0.407	0.050	0.138	0.138	0.407

Table 10.7 Results

Utility	A_1	A_2	A_3	A_4	A_5	A_6	A_7	A_8	A_9
Risk neutral	0.508	0.515	0.660	0.513	0.759	0.485	0.416	0.580	0.563
Risk aversion	0.635	0.672	0.878	0.787	0.887	0.650	0.641	0.808	0.762
Risk seeking	0.410	0.388	0.404	0.286	0.534	0.365	0.279	0.358	0.373

adding the utilities of the devices on different criteria as follows:

$$\text{Utility} = \sum_{1}^{30} W_j u_{ij} \tag{10.4}$$

The results, ranking the alternatives considered, are shown in Table 10.7. The utility of the LPG stove alternatives (A_5) is found to be the highest, followed by the Kerosene stove (A_3), for the risk-neutral and risk-aversion situations and by the Chulha (A_1) for the risk-seeking situation.

10.3 Concluding Remarks

Multi-attribute utility theory provides one such approach and is specially useful as it allows the treatment of both quantitative and qualitative criteria. Applications of this method can be found in the selection of solar photovoltaic alternatives [4], energy policy making [5], environmental impact assessment [6] and electric power system expansion planning [7]. A more detailed discussion of Multi-attribute utility theory can be found in Bunn [8] and Keeney and Raiffa [9].

References

1. Pohekar SD, Ramachandran M (2006) Multi-criteria evaluation of cooking devices with special reference to utility of parabolic solar cooker (PSC) in India. Energy 31:1215–1227
2. Bell DV, Keeney RL, Raiffa H (1978) Conflicting objectiws in decisions. International institute for applied systems analysis. Wiley, New York
3. Keeney R, Raiffa H (1993) Decisions with multiple objectives: preferences and value tradeoffs. Cambridge University Press, Cambridge.
4. Golabi K, Kirkwood CW, Sicherman A (1981) Selecting a portfolio of solar energy projects using Multi-attribute preference theory. Manage Sci 27(2):174–189
5. Jones M, Hope C, Hughes R (1990) A multi-attribute value model for the study of UK energy policy. J Oper Res Soc 41(10):919–929
6. McDaniels TL (1996) A multiattribute index for evaluating environmental impact of electric utilities. J Environ Manage 46:57–66
7. Voropai NL, Ivanova EY (2002) Multicriteria decision analysis technique in electric power system expansion planning. Int J Elec Power 24:71–78
8. Bunn D (1984). Applied decision analysis. McGraw-Hill, New York
9. Keeney R, Raiffa H (1976) Decision making with multiple objectives. Wiley, New York

Chapter 11
Fuzzy PROMETHEE

In many decision-making problems the decision-maker's judgments are not crisp, and it is relatively difficult for the decision-maker to provide precise numerical values for the criteria or attributes. These kinds of criteria make the evaluation process hard and vague. To deal with vagueness of human thought Zadeh [1] first introduced the fuzzy set theory, which was oriented to the rationality of uncertainty due to imprecision or vagueness. A major contribution of fuzzy set theory is its capability of representing vague data. In a classical set, an element belongs to, or does not belong to, a set whereas an element of a fuzzy set naturally belongs to the set with a membership value from the interval [0,1]. In this chapter, the procedure of the PROMETHEE method described in Chap. 5 will be applied making the assumption that the performance of alternative solutions are fuzzy while the preferences of the decision-maker, such as the parameters of generalized criteria and the weighting factors, are not. That is, the performance of alternative solutions can be determined only approximately and therefore is introduced into the calculations as a fuzzy number.

11.1 Fuzzy PROMETHEE

Fuzzy numbers can be presented, according to Dubois and Prade [2], in the form: $x = (m, a, b)_{LR}$. When the variable x has the value m it is certain to belong to the specific class and its membership function is $f(x) = 1$. If it has values smaller than $(m - a)$ and bigger than $(m + b)$ it does not belong in the specific class and $f(x) = 0$. In the interval between: $m - a < x < m + b$ there is a 'grade of membership' of x in the specific class associated with a number between 0 and 1. L and R are the functions indicating the change of $f(x)$ with x to the left and right of m, respectively. Assuming linear functions for L and R, the fuzzy number $x = (1, 0.1, 0.1)$ is presented in Fig. 11.1 as a triangle. The number 1.05 has a membership grade of 0.5 in the set.

J. R. San Cristóbal Mateo, *Multi-Criteria Analysis in the Renewable Energy Industry*, 73
Green Energy and Technology, DOI: 10.1007/978-1-4471-2346-0_11,
© Springer-Verlag London Limited 2012

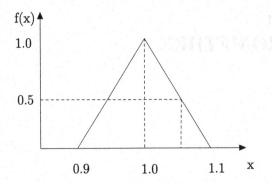

Fig. 11.1 Fuzzy number $x = (1, 0.1, 0.1)$

Table 11.1 Operations with fuzzy numbers $(m, a, b)_{LR}$

Addition	$(m, a, b)_{LR} \oplus (n, c, d)_{LR} = (m + n, a + c, b + d)_{LR}$
Opposite	$-(m, a, b)_{LR} = (-m, b, a)_{RL}$
Substraction	$(m, a, b)_{LR} - (n, c, d)_{RL} = (m - n, a + d, b + c)_{LR}$
Multiplication by scalar	$(m, a, b)_{LR} \times (n, 0, 0)_{LR} = (mn, an, bn)_{LR}$
Multiplication by fuzzy:	
- For $m > 0, n > 0$	$(m, a, b)_{LR} \otimes (n, c, d)_{LR} \approx (mn, cm + an, dm + bn)_{LR}$
- For $m < 0, n > 0$	$(m, a, b)_{RL} \otimes (n, c, d)_{LR} \approx (mn, an - dm, bn - cm)_{RL}$
- For $m < 0, n < 0$	$(m, a, b)_{LR} \otimes (n, c, d)_{LR} \approx (mn, -bn - dm, -an - cm)_{RL}$
Inverse (for $m > 0$)	$(m, a, b)_{LR}^{-1} \approx (m^{-1}, bm^{-2}, am^{-2})_{RL}$

The basic operations with fuzzy numbers used here are summarized in Table 11.1.

11.2 Application

Let us consider multi-criteria problem consisting in the selection of four alternative exploitation alternatives for the low enthalpy geothermal field presented in Chap. 5 [3]. Considering that the relevant data cannot be determined accurately, for the evaluation by the Fuzzy PROMETHEE method, the following statements are expressed as fuzzy inputs:

1. Economic evaluation. The net present value can deviate from the estimated value in the range of ± 10 percent for alternative 1; -15 percent to $+10$ percent for alternative 2; -18 percent to $+10$ percent for alternative 3 and -20 percent to $+10$ percent for alternative 4.
2. Jobs. The number of job positions can deviate by ± 2 for alternative 1 and by ± 4 for alternatives 2, 3 and 4.
3. Energy consumption. It can deviate by ± 20 percent from the estimated value.
4. Risk index. In the range of -1 percent to $+4$.

Table 11.2 Performance of alternatives expressed as fuzzy numbers [3]

Scenarios	NPV	Jobs	Energy use	Risk
A_1	(373,37,37)	(44,2,2)	(22.4,4.48,4.48)	(3,1,2)
A_2	(3706,556,371)	(99,4,4)	(51.0,10.2,10.2)	(6,1,2)
A_3	(3809,685,381))	(100,4,4)	(49.3,9,86,9,86)	(8,1,3)
A_4	(3860,772,386)	(104,4,4)	(47.5,9.5,9.5)	(9,1,4)

Table 11.3 Values of $\pi(A_i, A_k)$

	A_1	A_2	A_3	A_4
A_1	-	0.375	0.500	0.500
A_2	0.500	-	0.125	0.125
A_3	0.500	0.825	-	0
A_4	0.500	0.457	0.315	-

Table 11.4 Data for preorders P^+ and P^-

	A_1	A_2	A_3	A_4
$\phi^+(A_i)$	1.375	0.750	0.707	1.272
$\phi^-(A_k)$	1.500	0.915	0.940	0.750

According to the above estimations, the performance of the four alternatives expressed as fuzzy numbers is shown in Table 11.2.

To compare these fuzzy numbers the index proposed by Yager [4] will be used. The index corresponds to a 'weight average' of the fuzzy number and is determined by the center of weight of the surface representing its membership function:

$$F(m, a, b) = \frac{(3m - a + b)}{3} \tag{11.1}$$

Following step by step the same procedure of the PROMETHEE method described in Chap. 5 for the case where the performance of alternative solutions are considered with precise numerical values, we obtain for every couple of actions the values $\pi(A_i, A_k)$ shown in Table 11.3 and in Table 11.4 the outgoing and the incoming flow.

The intersection of the preorders is:

$$A_4 P^{(1)} A_2, \ A_4 P^{(1)} A_3, \ A_2 P^{(1)} A_3$$

Supposing that the decision-maker requests a total preorder, the net flows are calculated. Alternative A_4 is, again, the best ranked alternative. However, making the assumption that the performance of alternative solutions is fuzzy has changed the total preorder with regard to the PROMETHEE method as shown in Table 11.5. In this example criteria Jobs and Energy consumption have the same contribution to the outcome whether scalar numbers or fuzzy numbers are used because the Yager index of the fuzzy number coincides with m. The difference in performance arises from the two other criteria, Economic evaluation and Risk. The combined differences in these two criteria resulted in a change in the total ranking of the alternatives.

Table 11.5 Ranking

PROMETHEE	$A_4 > A_3 > A_1 > A_2$
F-PROMETHEE	$A_4 > A_1 > A_2 > A_3$

11.3 Concluding remarks

The Fuzzy PROMETHEE method is preferred when substantial uncertainties and subjectivities exist in certain environments and the vagueness of the linguistic terms in the evaluation process requires the employment of fuzzy numbers. In the literature, the method has been applied in environmental management for ranking contaminated sites [5], in logistic service selection [6] and material handling equipment selection [7, 8]. In the renewable energy industry, the applications are limited to the work by Goumas and Lygerou [3].

References

1. Zadeh LA (1965) Fussy sets. Inform Control 8:338–353
2. Dubois D, Prade H (1978) Operations on fuzzy numbers. Int J Syst Sci 9:613–626
3. Goumas M, Lygerou V (2000) An extension of the PROMETHEE method for decision making in fuzzy environment: ranking of alternative energy exploitation projects. Eur J Oper Res 123:606–613
4. Yager RR (1981) A procedure for ordering fuzzy subsets of the unit interval. Inform Sci 24:143–161
5. Zhang K, Kluck C, Achari G (2009) A comparative approach for ranking contaminated sites based on the risk assessment paradigm using fuzzy PROMETHEE. Environ Manage 44(5):952–967
6. Aloini D, Dulmin R, Mininno V (2010) A hybrid fuzzy-promethee method for logistics service selection. Int J Uncertain Fuzz 18(4):345–369
7. TuzkayaG Gülsün B, Kahraman C, Ozgen D (2010) An integrated fuzzy multi-criteria decision making methodology for material handling equipment selection problem and an application. Expert Syst Appl 37(4):2853–2863
8. Yilmaz B, Dagdeviren M (2011) A combined approach for equipment selection: F-PROMETHEE method and zero-one goal programming. Expert Syst Appl 38(9):11641–11650

Chapter 12
FAHP

The AHP method introduced by Saaty [1] shows the process of determining the priority of a set of alternatives and the relative importance of attributes in a multi-criteria decision-making problem. The primary advantage of the AHP approach is the relative ease with which it handles multiple criteria and performs qualitative and quantitative data. However, AHP is frequently criticized for its inability to adequately accommodate the inherent uncertainty and imprecision associated with mapping decision-maker perceptions to extract numbers [2].

Analysis of hierarchical structures in fuzzy environment was initially proposed by Buckley [3], who examined expressions of decision makers regarding with the pairwise comparisons while utilizing fuzzy ratios instead of crisp values. There are many alternatives as solution approaches [4, 5, 6] to perform on the Fuzzy AHP-based study on multiple criteria decision making problems. Buyukozkan et al. [7] introduced the main characteristics of different approaches on solving FAHP-based models by expressing their advantages and disadvantages with several points of view. In this chapter we use two methods in the selection of RE projects. The first one is a modified analytical hierarchy process proposed by Zeng et al. [8] while the second one is based on Chang's extent analysis [9, 10].

12.1 FAHP

The FAHP methodology consists of the following steps [8]:

Step 1. Measure factors in the hierarchy

The experts are required to provide their judgments on the basis of their knowledge and expertise for each factor at the bottom level in the hierarchy. Each expert can provide his or her judgments as a precise numerical value, a range of numerical values, a linguistic term or a fuzzy number. A score system is shown in Fig 12.1.

J. R. San Cristóbal Mateo, *Multi-Criteria Analysis in the Renewable Energy Industry*,
Green Energy and Technology, DOI: 10.1007/978-1-4471-2346-0_12,
© Springer-Verlag London Limited 2012

Fig. 12.1 Score system

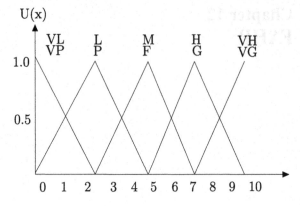

Fig. 12.2 Score system

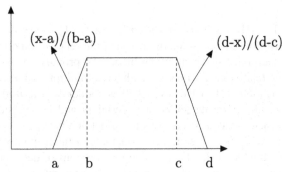

Step 2. Compare factors using pair-wise comparisons

The experts are required to compare every factor pair-wise in their corresponding section structured in the hierarchy and calibrate them on either a crisp or a fuzzy scale.

Step 3. Convert preferences into Standardized Trapezoidal Fuzzy Numbers (STFNs)

Because the values provided by experts are crisps, e.g., a numerical value, a range of numerical value, a linguistic term or a fuzzy number, the STFN is employed to convert these experts' judgments into a universal format for the composition of group preferences. A STFN can be defined as $\tilde{A} = (a, b, c, d)$, where $0 \leq a \leq b \leq c \leq d$ as shown in Fig. 12.2, and its membership function is defined as follows:

$$\mu_{\tilde{A}}(x) = \begin{cases} \frac{(x-a)}{(b-a)} & \text{for } a \leq x \leq b \\ 1 & \text{for } b \leq x \leq c \\ \frac{(d-x)}{(d-c)} & \text{for } c \leq x \leq d \\ 0 & \text{for otherwise} \end{cases} \qquad (12.1)$$

Step 4. Aggregate individual STFNs into group STFNs

The aim of this step is to apply an appropriate operator to aggregate individual preferences made by individual experts into a group preference of each factor. The aggregation of STFN scores is performed by applying the fuzzy weighted trapezoidal averaging operator, which is defined by:

$$\tilde{S} = \tilde{S}_{i1} \otimes c_1 \oplus \tilde{S}_{i2} \otimes c_2 \oplus \cdots \oplus \tilde{S}_{im} \otimes c_m \qquad (12.2)$$

where $\tilde{S}_i$ is the fuzzy aggregated score of the factor F_i, $\tilde{S}_{i1}, \tilde{S}_{i2}, \ldots, \tilde{S}_{im}$ are the STFN scores of the factor F_i measured by m experts $E_1, E_2, \ldots, E_m$ respectively, $\otimes$ and $\oplus$ denote the fuzzy multiplication operator and the fuzzy addition operator respectively, and $c_1, c_2, \ldots, c_m$ are contribution factors (CFs) allocated to experts, $E_1, E_2, \ldots, E_m$ and $c_1 + c_2 + \cdots + c_m = 1$. Similarly, the aggregation of STFN scales is defined as

$$\tilde{a} = \tilde{a}_{i1} \otimes c_1 \oplus \tilde{a}_{i2} \otimes c_2 \oplus \cdots \oplus \tilde{a}_{im} \otimes c_m \qquad (12.3)$$

where $\tilde{a}_{ij}$ is the aggregated fuzzy scale of F_i comparing to F_j; $i, j = 1, 2, \ldots, n$; $\tilde{a}_{ij1}$, $\tilde{a}_{ij2}, \ldots, \tilde{a}_{ijm}$ are the corresponding STFN scales of F_i comparing to F_j measured by experts $E_1, E_2, \ldots, E_m$, respectively.

Step 5. Defuzzify the SFTN scales

In order to convert the aggregated STFN scales into matching crisp values that can adequately represent the group preferences, a proper defuzzification method is needed. Assuming an aggregated STFN scale $\tilde{a}_{ij} = (a_{ij}^l, a_{ij}^m, a_{ij}^n, a_{ij}^u)$, the matching crisp value a_{ij} can be obtained

$$a_{ij} = \frac{a_{ij}^l + 2(a_{ij}^m + a_{ij}^n) + a_{ij}^u}{6} \qquad (12.4)$$

where $a_{ij} = 1$, $a_{ji} = \frac{1}{a_{ij}}$. Consequently, all the aggregated fuzzy scales $\tilde{a}_{ij}(i, j = 1, 2, \ldots, n)$ are transferred into crisp scales a_{ij} within the range $[0, 9]$.

Step 6. Calculate the priority of weights factors

Let $F_1, F_2, \ldots, F_n$ be a set of factors in one section, a_{ij} is the defuzzified scale representing the quantified judgment on F_i comparing to F_j. Pair-wise comparisons between F_i and F_j in the same section thus yields a n-by-n matrix defined as follows:

$$A = a_{ij} = \begin{array}{c} \\ F_1 \\ F_2 \\ \cdots \\ F_n \end{array} \begin{array}{cccc} F_1 & F_2 & \cdots & F_n \\ 1 & a_{12} & \cdots & a_{1n} \\ \frac{1}{a_{12}} & 1 & \cdots & a_{2n} \\ \cdots & \cdots & \cdots & \cdots \\ \frac{1}{a_{1n}} & \frac{1}{a_{2n}} & \cdots & 1 \end{array} \qquad (12.5)$$

where $a_{ij} = 1$, $a_{ji} = \frac{1}{a_{ij}}$ and $i, j = 1, 2, \ldots, n$.

The priority weights of factors in the matrix A can be calculated by using the arithmetic averaging method

$$w_i = \frac{1}{n} \sum_{j=1}^{n} \frac{a_{ij}}{\sum_{k=1}^{n} a_{kj}} \tag{12.6}$$

where w_i is the section weight of F_i. Assume F_i has t upper sections at different levels in the hierarchy, and $W_{\text{section}}^{(i)}$ is the section weight of the ith upper section which contains F_i in the hierarchy. The final weight w_i' of F_i can be derived by

$$w_i' = w_i \times \prod_{i=1}^{t} W_{\text{section}}^{(i)} \tag{12.7}$$

All individual upper section weights of $W_{\text{section}}^{(i)}$ can also be derived by Eq. 12.6 to prioritize sections within the corresponding cluster in the hierarchy.

Step 7. Calculate the fuzzy scores

When the scores and the priority weights of factors are obtained, the final fuzzy scores $(\widetilde{FS})$ can be calculated by

$$\widetilde{FS} = \sum_{i=1}^{n} \tilde{S} w_i' \tag{12.8}$$

$i = 1, 2, \ldots, n$

Step 8. Compare the $(\widetilde{FS})$ values using an outranking method

Several techniques have been proposed to rank final fuzzy scores [11]. In this case the approach proposed by Cheng and Cheng [12] will be used. Let A and B be two fuzzy numbers defined as follows:

$$f_A(x) = \begin{cases} f_A^L(x), & x < m_A \\ f_A^R(x), & x \geq m_A \end{cases} \tag{12.9}$$

$$f_B(x) = \begin{cases} f_B^L(x), & x < m_B \\ f_B^R(x), & x \geq m_B \end{cases} \tag{12.10}$$

where m_A and m_B are the mean of A and B. The metric distance between A and B can be calculated as follows:

$$D(A, B) = [\int_0^1 (g_A^L(y) - g_B^L(y))^2 dy + \int_0^1 (g_A^R(y) - g_B^R(y))^2 dy]^{\frac{1}{2}} \tag{12.11}$$

Table 12.1 Criteria and subcriteria

Criteria	Subcriteria
c_1 Technological	c_{11} Efficiency
	c_{12} Reliability
	c_{13} Maturity
c_2 Environmental	c_{21} Emissions avoided
	c_{22} Noise
	c_{23} Land use
c_3 Economic	c_{31} Investment costs
	c_{32} Operation and Maintenance costs
	c_{33} Service life

where g_A^L, g_A^R, g_B^L, and g_B^R are the inverse functions of f_A^L, f_A^R, f_B^L, and f_B^R, respectively.

In order to rank fuzzy numbers, let the fuzzy number $B = 0$, then the metric distance between A and 0 is calculated as follows:

$$D(A, 0) = [\int_0^1 (g_A^L(y))^2 dy + \int_0^1 (g_A^R(y))^2 dy]^{\frac{1}{2}} \tag{12.12}$$

the larger the value of $D(A,0)$, the better the ranking of A.

According to Cheng and Cheng [12], a trapezoidal fuzzy number $A = (a_1, a_2, a_3, a_4)$ can be approximated as a symmetry fuzzy number $S[\mu, \sigma]$, where μ denotes the mean of A, and σ denotes the standard deviation of A. The membership function of A is calculated as follows:

$$f_A(x) = \begin{cases} \frac{x-(\mu-\sigma)}{\sigma}, & \text{if } \mu - \sigma \leq x \leq \mu \\ \frac{(\mu+\sigma)-x}{\sigma}, & \text{if } \mu \leq x \leq \mu + \sigma \end{cases} \tag{12.13}$$

where μ and σ are calculated as follows:

$$\sigma = \frac{2(a_4 - a_1) + a_3 - a_2}{4} \tag{12.14}$$

$$\mu = \frac{a_1 + a_2 + a_3 + a_4}{4} \tag{12.15}$$

if $a_2 = a_3$, then A becomes a triangular fuzzy number, where $A = (a_1, a_2, a_4)$ and μ and σ can be calculated as follows:

$$\sigma = \frac{(a_4 - a_1)}{2} \tag{12.16}$$

$$\mu = \frac{a_1 + 2a_2 + a_4}{4} \tag{12.17}$$

The inverse functions g_A^L and g_A^R, of f_A^L and f_A^R respectively are shown as follows:

$$g_A^L(y) = (\mu - \sigma) + \sigma * y \tag{12.18}$$

$$g_A^R(y) = (\mu + \sigma) - \sigma * y \tag{12.19}$$

12.1.1 Application

Let us consider a multi-criteria problem consisting in the selection of a renewable energy project for electric generation. Five alternatives are eligible, Biomass (A_1), Geothermal (A_2), Hydropower (A_3), Solar (A_4) and Wind (A_5), which will be evaluated according to three decision criteria and nine sub-criteria as shown in Table 12.1.

Four decision-makers will evaluate the considered criteria to determine the most appropriate renewable energy alternative. Each decision-maker provides a decision about his/her judgment as a precise numerical value, a possible range of numerical value, a linguistic term or a fuzzy number. A score system is shown in Fig. 12.2.

Table 12.2 presents the decision-makers' opinions based on alternatives with respect to the related criteria. Then these values are converted into STFNs as defined in Eq. 12.1.

Table 12.3 presents the scores and converted STFN for alternative A_1. The aggregations of the obtained scores, calculated by Eq. 12.2 for the rest of alternatives are shown in Table 12.4.

The pair-wise comparisons of Technological, Environmental and Economic criteria and the corresponding STFNs are shown in Tables 12.5, 12.6, 12.7. The aggregation of STFNs scales are calculated by Eq. 12.3.

Then, STFN scale of comparisons are defuzzied by using Eq. 12.4 and by using Eq. 12.5, the pair-wise comparisons matrices are obtained. By taking into account these pair-wise comparisons matrices and using Eq. 12.6, the weights of the sub-criteria are obtained. Table 12.8 shows the the final weights of the criteria calculated using Eq. 12.7. The final Fuzzy scores obtained by using Eq 12.8, are shown in Table 12.9.

In the final step, the fuzzy scores are ranked using the method proposed by Cheng and Cheng [12]. The ranking results are shown in Table 12.10. The ranking of the alternatives is as follows: A_5, A_4, A_2, A_1 and A_3. According to this ranking, alternative A_5 (Wind) the most appropriate alternative.

12.2 FAHP Chang's Extent Analysis

One of the latest approaches on solution processes of FAHP methodology is based on Chang's extent analysis [9, 10]. The Chang's extent analysis is relatively easy while comparing the other approaches on FAHP. The fundamentals of the method

Table 12.2 Decision makers' assessments

	Biomass				Geothermal				Hydropower				Solar				Wind			
	DM_1	DM_2	DM_3	DM_4	DM_1	DM_2	DM_3	DM_4	DM_1	DM_2	DM_3	DM_4	DM_1	DM_2	DM_3	DM_4	DM_1	DM_2	DM_3	DM_4
C_{11}	G	8	7	8	G	8	7a8	A7	G	8	7a8	A7	8	7	4a5	8a9	9a10	8	8	9
C_{12}	7a8	7	8	7a8	7	8a9	8a9	7a8	8	7	7a8	6	7	8	5a6	8	A5	2	2a3	G
C_{13}	8	8	7	A7	8	8	8	8	8	8	6a7	7	7	7	7	7	G	4	4a5	A7
C_{21}	G	7	8a9	A8	G	8	8	A5	L	5	6a7	A8	G	8	8a9	8	8a9	VL	9	8
C_{22}	8	G	G	7a8	VG	8	7a8	6	G	6	7	8	G	9	9	A9	9	8	VL	8
C_{23}	G	7	8	8	G	7	7a8	6	L	6	6	7a8	G	8	8a9	9	A9	8a9	9	VL
C_{31}	4	G	7	8	G	9	8a9	A7	G	7	6	A7	7	8a9	8	7	8	8a9	8	VL
C_{32}	5	G	8	7	G	8	7a8	7	VG	8	7	7	8	8	7	8	8a9	VL	8	7a8
C_{33}	4	G	7	7	VG	8	8	8	G	7	7	7a8	7	8a9	8	8	A9	8a9	9	VL

Table 12.3 Scores and converted STFN for Biomass

	DM_1	DM_2	DM_3	DM_4
c_{11}	(5 7.5 7.5 10)	(8 8 8 8)	(7 7 7 7)	(8 8 8 8)
c_{12}	(7 7 8 8)	(7 7 7 7)	(8 8 8 8)	(7 7 8 8)
c_{13}	(8 8 8 8)	(8 8 8 8)	(7 7 7 7)	(6 7 7 8)
c_{21}	(5 7.5 7.5 10)	(7 7 7 7)	(8 8 9 9)	(7 8 8 9)
c_{22}	(8 8 8 8)	(5 7.5 7.5 10)	(5 7.5 7.5 10)	(7 7 8 8)
c_{23}	(5 7.5 7.5 10)	(7 7 7 7)	(8 8 8 8)	(8 8 8 8)
c_{31}	(4 4 4 4)	(5 7.5 7.5 10)	(7 7 7 7)	(8 8 8 8)
c_{32}	(5 5 5 5)	(5 7.5 7.5 10)	(8 8 8 8)	(7 7 7 7)
c_{33}	(4 4 4 4)	(5 7.5 7.5 10)	(7 7 7 7)	(7 7 7 7)

Table 12.4 Aggregated STFNs

	Biomass	Geothermal	Hydropower	Solar	Wind
c_{11}	(7 7.6 7.6 8.2)	(6.5 7.4 7.6 8.5)	(6.5 7.4 7.6 8.5)	(6.7 6.7 7.2 7.2)	(8.5 8.5 8.7 8.7)
c_{12}	(7.2 7.2 7.7 7.7)	(7.2 7.2 7.7 7.7)	(7 7 7.2 7.2)	(7 7 7.2 7.2)	(3.2 4.1 4.4 5.2)
c_{13}	(7.2 7.5 7.5 7.7)	(8 8 8 8)	(7.2 7.2 7.5 7.5)	(7 7 7 7)	(4.7 5.6 5.9 6.7)
c_{21}	(6.7 7.6 7.9 8.7)	(6.2 7.1 7.1 8)	(4.5 4.7 5 5.2)	(7.2 7.9 8.1 8.7)	(8.1 8.7 9 9)
c_{22}	(6.2 7.5 7.7 9)	(7.1 7.7 8 8)	(6.5 7.1 7.1 7.7)	(7.7 8.6 8.6 9.5)	(8.1 8.7 8.7 8.7)
c_{23}	(7 7.6 7.6 8.2)	(6.2 6.9 7.1 7.7)	(4.7 4.7 5 5)	(7.5 8.1 8.4 9)	(8.1 9 9.2 9.5)
c_{31}	(6 6.6 6.6 7.2)	(7 7.9 8.1 9)	(6.2 7.1 7.1 8)	(7.5 7.5 7.7 7.7)	(7.9 8.5 8.7 8.7)
c_{32}	(6.2 6.9 6.9 7.5)	(6.7 7.4 7.6 8.2)	(5.2 5.2 5.2 5.2)	(7.7 7.7 7.7 7.7)	(7.6 8.2 8.7 8.7)
c_{33}	(5.7 6.4 6.4 7)	(7.9 8.5 8.5 8.5)	(6.5 7.1 7.4 8)	(7.5 7.78 8)	(8.1 9 9.2 9.5)

can be examined in four steps: Obtaining priority weights, Comparing degrees of possibility, Obtaining the weight vector and Ranking of the alternatives:

Step 1. Obtaining priority weights

The first step in this method is to use Fuzzy triangular numbers for pairwise comparison by means of FAHP scale and next to use extent analysis method to obtain priority weights by using synthetic extent values. The value of fuzzy synthetic extent with respect to the ith object is represented as:

$$S_i = \sum_{j=1}^{m} M_{gi}^j \otimes [\sum_{i=1}^{n} \sum_{j=1}^{m} M_{gi}^j]^{-1} \qquad (12.20)$$

and fuzzy addition operation of m extent analysis values can be performed for particular matrix such that

$$\sum_{j=1}^{m} M_{gi}^j = (\sum_{j=1}^{m} l_j, \sum_{j=1}^{m} m_j, \sum_{j=1}^{m} u_j) \qquad (12.21)$$

to obtain $\sum_{j=1}^{m} M_{gi}^j$.

Table 12.5 Fuzzy aggregation of technological criteria

	c_{11}		c_{12}		c_{13}	
	Scale	Converted SFTN	Scale	Converted SFTN	Scale	Converted SFTN
DM_1			3 a 4	(3 3 4 4)	3	(3 3 3 3)
DM_2			5	(5 5 5 5)	2 a 3	(2 2 3 3)
DM_3			5	(5 5 5 5)	4	(4 4 4 4)
DM_4			4	(4 4 4 4)	3 a 4	(3 3 4 4)
Aggregation		(1 1 1 1)		(4.3 4.3 4.5 4.5)		(3 3 3.5 3.5)
DM_1					6	(6 6 6 6)
DM_2					5 a 6	(5 5 6 6)
DM_3					6	(6 6 6 6)
DM_4					4 a 5	(4 4 5 5)
Aggregation				(1 1 1 1)		(5.3 5.3 5.8 5.8)
DM_1						
DM_2						
DM_3						
DM_4						
Aggregation						(1 1 1 1)

Table 12.6 Fuzzy aggregation of environmental criteria

	c_{21}		c_{22}		c_{23}	
	Scale	converted SFTN	Scale	converted SFTN	Scale	convertedSFTN
DM_1			6 a 7	(6 6 7 7)	1 a 2	(1 1 2 2)
DM_2			7 a 8	(7 7 8 8)	1	(1 1 1 1)
DM_3			6 a 7	(6 6 7 7)	1 a 2	(1 1 2 2)
DM_4			4	(4 4 4 4)	3 a 4	(3 3 4 4)
Aggregation		(1 1 1 1)		(6.3 6.25 7.25 7.3)		(1 1 1.75 1.8)
DM_1					2	(2 2 2 2)
DM_2					2 a 3	(2 2 3 3)
DM_3					1	(1 1 1 1)
DM_4					2 a 3	(2 2 3 3)
Aggregation				(1 1 1 1)		(1.8 1.8 2.2 2.3)
DM_1						
DM_2						
DM_3						
DM_4						
Aggregation						(1 1 1 1)

Then, the fuzzy addition operation of $M_{gi}^j (j = 1, 2, \ldots, m)$ values such that

$$\sum_{i=1}^{n} \sum_{j=1}^{m} M_{gi}^j = \left(\sum_{i=1}^{n} l_i, \sum_{i=1}^{n} m_i, \sum_{i=1}^{n} u_i \right) \qquad (12.22)$$

Table 12.7 Fuzzy aggregation of economic criteria

	c_{31}		c_{32}		c_{33}	
	Scale	Converted SFTN	Scale	Converted SFTN	Scale	Converted SFTN
DM_1			3 a 4	(3 3 4 4)	1 a 2	(1 1 2 2)
DM_2			4	(4 4 4 4)	1	(1 1 1 1)
DM_3			3	(3 3 3 3)	2	(2 2 2 2)
DM_4			4	(4 4 4 4)	1 a 2	(1 1 2 2)
Aggregation		(1 1 1 1)		(3.5 3.5 3.7 3.8)		(1.3 1.3 1.7 1.8)
DM_1					5 a 6	(5 5 6 6)
DM_2					5 a 5	(5 5 5 5)
DM_3					5	(5 5 5 5)
DM_4					4 a 5	(4 4 5 5)
Aggregation				(1 1 1 1)		(4.8 4.8 5.25 5.3)
DM_1						
DM_2						
DM_3						
DM_4						
Aggregation						(1 1 1 1)

Table 12.8 Weight vectors

Technological	[0.06	0.03	0.01]
Environmental	[0.34	0.13	0.14]
Economic	[0.14	0.10	0.05]

Table 12.9 Fuzzy scores

Alternative	Fuzzy Score
A_1	(6.5 7.3 7.4 8.2)
A_2	(6.7 7.4 7.6 8.2)
A_3	(5.4 5.8 6 6.3)
A_4	(7.4 7.8 8 8.5)
A_5	(7.9 8.5 8.8 8.5)

Table 12.10 Ranking of alternatives

Alternative	$D(A, 0)$	Ranking
A_1	10.47	4
A_2	10.52	3
A_3	8.34	5
A_4	11.24	2
A_5	12.04	1

are performed to obtain $[\sum_{i=1}^{n} \sum_{j=1}^{m} M_{gi}^{j}]^{-1}$

At the end of Step 1, the inverse of the determined vector can be expressed as follows:

$$[\sum_{i=1}^{n} \sum_{j=1}^{m} M_{gi}^{j}]^{-1} = (\frac{1}{\sum_{i=1}^{n} u_i} \frac{1}{\sum_{i=1}^{n} m_i} \frac{1}{\sum_{i=1}^{n} l_i}) \qquad (12.23)$$

Table 12.11 Alternatives and criteria

Alternatives
A_1 Wind power $10 \leq P \leq 50Mw$
A_2 Hydroelectric $25 \leq P \leq 50Mw$
A_3 Solar thermo-electric $P \geq 10Mw$
A_4 Biomass (energetic cultivations) $P \leq 5Mw$
A_5 Biomass (forest and agricultural wastes) $P \leq 5Mw$
A_6 Biomass (Co-combustion in conventional central) $P \geq 50Mw$
A_7 Biogas (electric generation) $P \leq 2Mw$
Criteria
Power $(P)(Mw)$
Investment ratio (IR)
Operating hours (OH)
Useful life (UL)
Operation and maintenance costs (OM)
Emissions avoided $(tCo2)$

Table 12.12 FAHP scale

Definition	Intensity of importance
Equal	$(1,1,1)$
Weak	$(\frac{2}{3}, 1, \frac{3}{2})$
Fairly strong	$(\frac{3}{2}, 2, \frac{2}{5})$
Very string	$(\frac{5}{2}, 3, \frac{7}{2})$
Absolute	$(\frac{7}{2}, 4, \frac{9}{2})$

Step 2. Comparing degrees of possibility

The degree of possibility of $M_2 = (l_2, m_2, u_2) \geq M_1 = (l_1, m_1, u_1)$ is defined as

$$V(M_2 \geq M_1) = sup[min(\lambda\pi_1(x), \lambda\pi_2(y))]_{y \geq x} \qquad (12.24)$$

where x and y, are the values on the axis of membership function of each criterion and it can be represented as follows:

$$V(M_2 \geq M_1) = hgt(M_1 \cap M_2) = \lambda\pi_2(d) = \begin{cases} 1 & \text{for } m_2 \geq m_1 \\ 0 & \text{for } l_2 \geq l_1 \\ \frac{(l_1 - u_2)}{(m_2 - u_2) - (m_1 - u1)} & \text{otherwise} \end{cases}$$
$$(12.25)$$

where d is the ordinate of the highest intersection point between $\mu\pi_1$ and $\mu\pi_2$. To be able to compare we need both the values of $V(M_1 \geq M_2)$ and $V(M_2 \geq M_1)$.

Step 3. Obtaining the weight vector

The degree possibility for a convex fuzzy number to be greater than k convex fuzzy numbers $M_i (i = 1, 2, k)$ can be defined by

Table 12.13 Evaluation of RE project choice criteria with respect to goal

	P	IR	OH	UL	OM	tco2
P	(1,1,1)	$(\frac{2}{5},3,\frac{7}{2})$	$(\frac{2}{5},\frac{1}{2},\frac{2}{3})$	$(\frac{2}{3},1,\frac{3}{2})$	$(\frac{2}{7},\frac{1}{3},\frac{2}{5})$	$(\frac{2}{5},\frac{1}{2},\frac{2}{3})$
IR	$(\frac{2}{7},\frac{1}{3},\frac{2}{5})$	(1,1,1)	$(\frac{2}{7},\frac{1}{3},\frac{2}{5})$	$(\frac{2}{5},\frac{1}{2},\frac{2}{3})$	(1,1,1)	$(\frac{2}{5},1,\frac{3}{2})$
OH	$(\frac{3}{2},2,\frac{5}{2})$	$(\frac{2}{5},3,\frac{7}{2})$	(1,1,1)	$(\frac{2}{5},\frac{1}{2},\frac{2}{3})$	$(\frac{2}{3},1,\frac{3}{2})$	$(\frac{2}{7},\frac{1}{3},\frac{2}{5})$
UL	$(\frac{2}{3},1,\frac{3}{2})$	$(\frac{3}{2},2,\frac{5}{2})$	$(\frac{3}{2},2,\frac{5}{2})$	(1,1,1)	$(\frac{2}{9},\frac{1}{4},\frac{2}{7})$	(1,1,1)
OM	$(\frac{3}{2},3,\frac{7}{2})$	(1,1,1)	$(\frac{2}{5},1,\frac{3}{2})$	$(\frac{7}{2},4,\frac{9}{2})$	(1,1,1)	$(\frac{2}{5},\frac{1}{2},\frac{2}{3})$
tco2	$(\frac{3}{2},2,\frac{5}{2})$	$(\frac{2}{3},1,\frac{3}{2})$	$(\frac{2}{3},3,\frac{7}{2})$	(1,1,1)	$(\frac{3}{2},2,\frac{5}{2})$	(1,1,1)

$$V(M \geq M_1, M_2, \ldots, M_k) = V[(M \geq M_1), (M \geq M_2), \ldots, (M \geq M_k)] \tag{12.26}$$
$$min V(M \geq M_i)$$

$i = 1, 2, \ldots, k$

Assume that, $d^t(A_i) = min V(S_i \geq S_k)$ for $k = 1, 2, \ldots, m; k \neq i$. Then, the weight vector is given by

$$W^t = (d^t(A_1), d^t(A_2), d^t(A_n))^T \tag{12.27}$$

where $A_i (i = 1, 2, \ldots, n)$ are n elements. Via normalization, the normalized weight vector is:

$$W = (d(A_1), d(A_2), \ldots, d(A_n))^T \tag{12.28}$$

where W is a nonfuzzy number that gives priority weights of an attribute or an alternative over other.

Step 4. Ranking of the alternatives

Based on the weights of the alternatives with respect to the main goal, the ranking of the different alternatives is obtained

12.2.1 Application

In this subsection the FAHP model is applied to a renewable energy project selection for electric power generation. Seven alternatives are available which will be evaluated according to six criteria as shown in Table 12.11. The evaluation scale used by experts is shown in Table 12.12. The criteria considered are compared with respect to goal in Table 12.13, and comparative evaluations of the alternatives with respect to criteria are shown in Tables 12.14, 12.15, 12.16, 12.17, 12.18, 12.19.

At the end of Step 1, the normalized weight vector obtained from Table 12.13 is

Table 12.14 Evaluation of RE project choice criteria with respect to Power

	A_1	A_2	A_3	A_4	A_5	A_6	A_7
A_1	$(1,1,1)$	$(\frac{2}{3},1,\frac{3}{2})$	$(\frac{2}{3},1,\frac{3}{2})$	$(\frac{5}{3},3,\frac{7}{2})$	$(\frac{5}{3},3,\frac{7}{2})$	$(\frac{2}{3},1,\frac{3}{2})$	$(\frac{7}{2},4,\frac{9}{2})$
A_2	$(\frac{2}{3},1,\frac{3}{2})$	$(1,1,1)$	$(\frac{2}{3},1,\frac{3}{2})$	$(\frac{5}{3},3,\frac{7}{2})$	$(\frac{5}{3},3,\frac{7}{2})$	$(1,1,1)$	$(\frac{7}{2},4,\frac{9}{2})$
A_3	$(\frac{2}{3},1,\frac{3}{2})$	$(\frac{2}{3},1,\frac{3}{2})$	$(1,1,1)$	$(\frac{5}{2},3,\frac{7}{2})$	$(\frac{5}{2},3,\frac{7}{2})$	$(1,1,1)$	$(\frac{7}{2},4,\frac{9}{2})$
A_4	$(\frac{2}{7},\frac{1}{3},\frac{2}{5})$	$(\frac{2}{7},\frac{1}{3},\frac{2}{5})$	$(\frac{2}{7},\frac{1}{3},\frac{2}{5})$	$(1,1,1)$	$(1,1,1)$	$(\frac{2}{9},\frac{1}{4},\frac{2}{7})$	$(1,1,1)$
A_5	$(\frac{2}{7},\frac{1}{3},\frac{2}{5})$	$(\frac{2}{7},\frac{1}{3},\frac{2}{5})$	$(\frac{2}{7},\frac{1}{3},\frac{2}{5})$	$(1,1,1)$	$(1,1,1)$	$(\frac{2}{9},\frac{1}{4},\frac{2}{7})$	$(1,1,1)$
A_6	$(\frac{2}{3},1,\frac{3}{2})$	$(1,1,1)$	$(1,1,1)$	$(\frac{7}{2},4,\frac{9}{2})$	$(\frac{7}{2},4,\frac{9}{2})$	$(1,1,1)$	$(\frac{7}{2},4,\frac{9}{2})$
A_7	$(\frac{2}{9},\frac{1}{4},\frac{2}{7})$	$(\frac{2}{9},\frac{1}{4},\frac{2}{7})$	$(\frac{2}{9},\frac{1}{4},\frac{2}{7})$	$(1,1,1)$	$(1,1,1)$	$(\frac{2}{9},\frac{1}{4},\frac{2}{7})$	$(1,1,1)$

Table 12.15 Evaluation of RE project choice criteria with respect to Investment Ratio

	A_1	A_2	A_3	A_4	A_5	A_6	A_7
A_1	$(1,1,1)$	$(\frac{2}{5},\frac{1}{2},\frac{2}{3})$	$(\frac{7}{2},4,\frac{9}{2})$	$(\frac{5}{2},3,\frac{7}{2})$	$(\frac{5}{2},3,\frac{7}{2})$	$(1,1,1)$	$(1,1,1)$
A_2	$(\frac{3}{2},2,\frac{5}{2})$	$(1,1,1)$	$(\frac{7}{2},4,\frac{9}{2})$	$(\frac{5}{2},3,\frac{7}{2})$	$(\frac{5}{2},3,\frac{7}{2})$	$(\frac{3}{2},2,\frac{5}{2})$	$(\frac{3}{2},2,\frac{5}{2})$
A_3	$(\frac{2}{9},\frac{1}{4},\frac{2}{7})$	$(\frac{2}{9},\frac{1}{4},\frac{2}{7})$	$(1,1,1)$	$(\frac{2}{7},\frac{1}{3},\frac{2}{5})$	$(\frac{2}{7},\frac{1}{3},\frac{2}{5})$	$(\frac{2}{9},\frac{1}{4},\frac{2}{7})$	$(\frac{3}{2},2,\frac{5}{2})$
A_4	$(\frac{2}{7},\frac{1}{3},\frac{2}{5})$	$(\frac{2}{7},\frac{1}{3},\frac{2}{5})$	$(\frac{5}{2},3,\frac{7}{2})$	$(1,1,1)$	$(1,1,1)$	$(\frac{2}{7},\frac{1}{3},\frac{2}{5})$	$(1,1,1)$
A_5	$(\frac{2}{7},\frac{1}{3},\frac{2}{5})$	$(\frac{2}{7},\frac{1}{3},\frac{2}{5})$	$(\frac{5}{2},3,\frac{7}{2})$	$(1,1,1)$	$(1,1,1)$	$(\frac{2}{7},\frac{1}{3},\frac{2}{5})$	$(1,1,1)$
A_6	$(1,1,1)$	$(\frac{2}{5},\frac{1}{2},\frac{2}{3})$	$(\frac{7}{2},4,\frac{9}{2})$	$(\frac{5}{2},3,\frac{7}{2})$	$(\frac{5}{2},3,\frac{7}{2})$	$(1,1,1)$	$(1,1,1)$
A_7	$(1,1,1)$	$(\frac{2}{5},\frac{1}{2},\frac{2}{3})$	$(\frac{2}{5},\frac{1}{2},\frac{2}{3})$	$(1,1,1)$	$(1,1,1)$	$(1,1,1)$	$(1,1,1)$

Table 12.16 Evaluation of RE project choice criteria with respect to Operating Hours

	A_1	A_2	A_3	A_4	A_5	A_6	A_7
A_1	$(1,1,1)$	$(\frac{3}{2},2,\frac{5}{2})$	$(\frac{2}{5},\frac{1}{2},\frac{2}{3})$	$(\frac{2}{9},\frac{1}{4},\frac{2}{7})$	$(\frac{2}{9},\frac{1}{4},\frac{2}{7})$	$(\frac{2}{9},\frac{1}{4},\frac{2}{7})$	$(\frac{2}{9},\frac{1}{4},\frac{2}{7})$
A_2	$(\frac{2}{5},\frac{1}{2},\frac{2}{3})$	$(1,1,1)$	$(\frac{2}{5},\frac{1}{2},\frac{2}{3})$	$(\frac{2}{9},\frac{1}{4},\frac{2}{7})$	$(\frac{2}{9},\frac{1}{4},\frac{2}{7})$	$(\frac{2}{9},\frac{1}{4},\frac{2}{7})$	$(\frac{2}{9},\frac{1}{4},\frac{2}{7})$
A_3	$(\frac{3}{2},2,\frac{5}{2})$	$(\frac{3}{2},2,\frac{5}{2})$	$(1,1,1)$	$(\frac{2}{9},\frac{1}{4},\frac{2}{7})$	$(\frac{2}{9},\frac{1}{4},\frac{2}{7})$	$(\frac{2}{9},\frac{1}{4},\frac{2}{7})$	$(\frac{2}{7},\frac{1}{3},\frac{2}{5})$
A_4	$(\frac{7}{2},4,\frac{9}{2})$	$(\frac{7}{2},4,\frac{9}{2})$	$(\frac{7}{2},4,\frac{9}{2})$	$(1,1,1)$	$(1,1,1)$	$(1,1,1)$	$(1,1,1)$
A_5	$(\frac{7}{2},4,\frac{9}{2})$	$(\frac{7}{2},4,\frac{9}{2})$	$(\frac{7}{2},4,\frac{9}{2})$	$(1,1,1)$	$(1,1,1)$	$(1,1,1)$	$(1,1,1)$
A_6	$(\frac{7}{2},4,\frac{9}{2})$	$(\frac{7}{2},4,\frac{9}{2})$	$(\frac{7}{2},4,\frac{9}{2})$	$(1,1,1)$	$(1,1,1)$	$(1,1,1)$	$(1,1,1)$
A_4	$(\frac{7}{2},4,\frac{9}{2})$	$(\frac{7}{2},4,\frac{9}{2})$	$(\frac{5}{2},3,\frac{7}{2})$	$(1,1,1)$	$(1,1,1)$	$(1,1,1)$	$(1,1,1)$

Table 12.17 Evaluation of RE project choice criteria with respect to Useful Life

	A_1	A_2	A_3	A_4	A_5	A_6	A_7
A_1	$(1,1,1)$	$(\frac{2}{5},\frac{1}{2},\frac{2}{3})$	$(\frac{2}{5},\frac{1}{2},\frac{2}{3})$	$(\frac{3}{2},2,\frac{5}{2})$	$(\frac{3}{2},2,\frac{5}{2})$	$(1,1,1)$	$(1,1,1)$
A_2	$(\frac{3}{2},2,\frac{5}{2})$	$(1,1,1)$	$(1,1,1)$	$(\frac{5}{2},3,\frac{7}{2})$	$(\frac{5}{2},3,\frac{7}{2})$	$(\frac{3}{2},2,\frac{5}{2})$	$(\frac{3}{2},2,\frac{5}{2})$
A_3	$(\frac{3}{2},2,\frac{5}{2})$	$(1,1,1)$	$(1,1,1)$	$(\frac{5}{2},3,\frac{7}{2})$	$(\frac{5}{2},3,\frac{7}{2})$	$(\frac{3}{2},2,\frac{5}{2})$	$(\frac{3}{2},2,\frac{5}{2})$
A_4	$(\frac{2}{5},\frac{1}{2},\frac{2}{3})$	$(\frac{2}{7},\frac{1}{3},\frac{2}{5})$	$(\frac{2}{7},\frac{1}{3},\frac{2}{5})$	$(1,1,1)$	$(1,1,1)$	$(\frac{2}{5},\frac{1}{2},\frac{2}{3})$	$(\frac{2}{5},\frac{1}{2},\frac{2}{3})$
A_5	$(\frac{2}{5},\frac{1}{2},\frac{2}{3})$	$(\frac{2}{7},\frac{1}{3},\frac{2}{5})$	$(\frac{2}{7},\frac{1}{3},\frac{2}{5})$	$(1,1,1)$	$(1,1,1)$	$(\frac{2}{5},\frac{1}{2},\frac{2}{3})$	$(\frac{2}{5},\frac{1}{2},\frac{2}{3})$
A_6	$(1,1,1)$	$(\frac{2}{5},\frac{1}{2},\frac{2}{3})$	$(\frac{2}{5},\frac{1}{2},\frac{2}{3})$	$(\frac{3}{2},2,\frac{5}{2})$	$(\frac{3}{2},2,\frac{5}{2})$	$(1,1,1)$	$(1,1,1)$
A_7	$(1,1,1)$	$(\frac{2}{5},\frac{1}{2},\frac{2}{3})$	$(\frac{2}{5},\frac{1}{2},\frac{2}{3})$	$(\frac{3}{2},2,\frac{5}{2})$	$(\frac{3}{2},2,\frac{5}{2})$	$(1,1,1)$	$(1,1,1)$

Table 12.18 Evaluation of RE project choice criteria with respect to Operation and Maintenance costs

	A_1	A_2	A_3	A_4	A_5	A_6	A_7
A_1	(1,1,1)	$(\frac{2}{5},\frac{1}{2},\frac{2}{3})$	$(\frac{5}{2},3,\frac{7}{2})$	$(\frac{2}{5},\frac{1}{2},\frac{2}{3})$	$(\frac{2}{5},\frac{1}{2},\frac{2}{3})$	$(\frac{2}{5},\frac{1}{2},\frac{2}{3})$	(1,1,1)
A_2	$(\frac{3}{2},2,\frac{5}{2})$	(1,1,1)	$(\frac{7}{2},4,\frac{9}{2})$	(1,1,1)	(1,1,1)	(1,1,1)	$(\frac{5}{2},3,\frac{7}{2})$
A_3	$(\frac{2}{7},\frac{1}{3},\frac{2}{5})$	$(\frac{2}{9},\frac{1}{4},\frac{2}{7})$	(1,1,1)	$(\frac{2}{9},\frac{1}{4},\frac{2}{7})$	$(\frac{2}{9},\frac{1}{4},\frac{2}{7})$	$(\frac{2}{9},\frac{1}{4},\frac{2}{7})$	$(\frac{2}{9},\frac{1}{4},\frac{2}{7})$
A_4	$(\frac{3}{2},2,\frac{5}{2})$	(1,1,1)	$(\frac{7}{2},4,\frac{9}{2})$	(1,1,1)	(1,1,1)	(1,1,1)	$(\frac{5}{2},3,\frac{7}{2})$
A_5	$(\frac{3}{2},2,\frac{5}{2})$	(1,1,1)	$(\frac{7}{2},4,\frac{9}{2})$	(1,1,1)	(1,1,1)	(1,1,1)	$(\frac{5}{2},3,\frac{7}{2})$
A_6	$(\frac{3}{2},2,\frac{5}{2})$	(1,1,1)	$(\frac{7}{2},4,\frac{9}{2})$	(1,1,1)	(1,1,1)	(1,1,1)	$(\frac{5}{2},3,\frac{7}{2})$
A_7	(1,1,1)	$(\frac{2}{7},\frac{1}{3},\frac{2}{5})$	$(\frac{5}{2},3,\frac{7}{2})$	$(\frac{2}{7},\frac{1}{3},\frac{2}{5})$	$(\frac{2}{7},\frac{1}{3},\frac{2}{5})$	$(\frac{2}{7},\frac{1}{3},\frac{2}{5})$	(1,1,1)

Table 12.19 Evaluation of RE project choice criteria with respect to Emissions avoided

	A_1	A_2	A_3	A_4	A_5	A_6	A_7
A_1	(1,1,1)	$(\frac{7}{2},4,\frac{9}{2})$	$(\frac{5}{2},3,\frac{7}{2})$	$(\frac{5}{2},3,\frac{7}{2})$	$(\frac{5}{2},3,\frac{7}{2})$	$(\frac{3}{2},2,\frac{5}{2})$	$(\frac{7}{2},4,\frac{9}{2})$
A_2	$(\frac{2}{9},\frac{1}{4},\frac{2}{7})$	(1,1,1)	$(\frac{2}{5},\frac{1}{2},\frac{2}{3})$	$(\frac{2}{7},\frac{1}{3},\frac{2}{5})$	$(\frac{2}{7},\frac{1}{3},\frac{2}{5})$	$(\frac{2}{9},\frac{1}{4},\frac{2}{7})$	(1,1,1)
A_3	$(\frac{2}{7},\frac{1}{3},\frac{2}{5})$	$(\frac{3}{2},2,\frac{5}{2})$	(1,1,1)	$(\frac{2}{7},\frac{1}{3},\frac{2}{5})$	$(\frac{2}{7},\frac{1}{3},\frac{2}{5})$	$(\frac{2}{9},\frac{1}{4},\frac{2}{7})$	$(\frac{3}{2},2,\frac{5}{2})$
A_4	$(\frac{2}{7},\frac{1}{3},\frac{2}{5})$	$(\frac{5}{2},3,\frac{7}{2})$	$(\frac{5}{2},3,\frac{7}{2})$	(1,1,1)	(1,1,1)	$(\frac{2}{5},\frac{1}{2},\frac{2}{3})$	$(\frac{5}{2},3,\frac{7}{2})$
A_5	$(\frac{2}{7},\frac{1}{3},\frac{2}{5})$	$(\frac{5}{2},3,\frac{7}{2})$	$(\frac{5}{2},3,\frac{7}{2})$	(1,1,1)	(1,1,1)	$(\frac{5}{2},3,\frac{7}{2})$	$(\frac{5}{2},3,\frac{7}{2})$
A_6	$(\frac{2}{5},\frac{1}{2},\frac{2}{3})$	$(\frac{7}{2},4,\frac{9}{2})$	$(\frac{7}{2},4,\frac{9}{2})$	$(\frac{5}{2},3,\frac{7}{2})$	$(\frac{5}{2},3,\frac{7}{2})$	(1,1,1)	$(\frac{5}{2},3,\frac{7}{2})$
A_7	$(\frac{2}{9},\frac{1}{4},\frac{2}{7})$	(1,1,1)	$(\frac{2}{5},\frac{1}{2},\frac{2}{3})$	$(\frac{2}{7},\frac{1}{3},\frac{2}{5})$	$(\frac{2}{7},\frac{1}{3},\frac{2}{5})$	$(\frac{2}{7},\frac{1}{3},\frac{2}{5})$	(1,1,1)

$$[\sum_{i=1}^{n}\sum_{j=1}^{m} M_{gi}^{j}]^{-1} = (\frac{1}{55.22}\,\frac{1}{46.08}\,\frac{1}{38.37})$$

In order to identify the computation stages clearly, the pair-wise comparisons from Table 12.13 are evaluated as follows:

$$S_P = (5.25, 6.33, 7.73) \otimes (\frac{1}{55.22}\,\frac{1}{46.08}\,\frac{1}{38.37}) = (0.10, 0.14, 0.20)$$

$$S_{IR} = (3.64, 4.17, 4.97) \otimes (\frac{1}{55.22}\,\frac{1}{46.08}\,\frac{1}{38.37}) = (0.07, 0.09, 0.13)$$

$$S_{OH} = (6.35, 7.83, 9.57) \otimes (\frac{1}{55.22}\,\frac{1}{46.08}\,\frac{1}{38.37}) = (0.12, 0.17, 0.25)$$

$$S_{UL} = (5.89, 7.25, 8.79) \otimes (\frac{1}{55.22}\,\frac{1}{46.08}\,\frac{1}{38.37}) = (0.11, 0.16, 0.23)$$

$$S_{OM} = (9.07, 10.5, 12.17) \otimes (\frac{1}{55.22}\,\frac{1}{46.08}\,\frac{1}{38.37}) = (0.16, 0.23, 0.32)$$

$$S_{tco2} = (8.17, 10.0, 12.0) \otimes (\frac{1}{55.22}\,\frac{1}{46.08}\,\frac{1}{38.37}) = (0.15, 0.22, 0.31)$$

After determining these results the following are obtained

Table 12.20 Weights of alternatives with respect to goal

	P (0.09)	IR (0)	OH (0.18)	UL (0.15)	OM (0.30)	tco2 (0.28)
A_1	0.22	0.27	0	0.07	0	0.39
A_2	0.22	0.46	0	0.40	0.25	0
A_3	0.22	0	0	0.40	0	0
A_4	0	0	0.25	0	0.25	0.18
A_5	0	0	0.25	0.07	0.25	0.26
A_6	0.33	0.27	0.25	0.07	0.25	0.26
A_7	0	0	0.25	0.07	0	0

$$V(S_P \geq S_{IR}) = 1$$
$$V(S_P \geq S_{OH}) = 0.72$$
$$V(S_P \geq S_{UL}) = 0.81$$
$$V(S_P \geq S_{OM}) = 0.31$$
$$V(S_P \geq S_{RE}) = 0.38$$

Applying Eq. 12.26:

$$\min(S_P \geq S_{IR}, S_{IP}, S_{OH}, S_{UL}, S_{OM}) = 0.31$$
$$\min(S_{IR} \geq S_P, S_{IP}, S_{OH}, S_{UL}, S_{OM}) = 0$$
$$\min(S_{IP} \geq S_P, S_{IR}, S_{OH}, S_{UL}, S_{OM}) = 0.60$$
$$\min(S_{OH} \geq S_P, S_{IR}, S_{IP}, S_{UL}, S_{OM}) = 0.50$$
$$\min(S_{UL} \geq S_P, S_{IR}, S_{IP}, S_{OH}, S_{OM}) = 1$$
$$\min(S_{OM} \geq S_P, S_{IR}, S_{IP}, S_{OH}, S_{UL}) = 0.94$$

Therefore, the weight vector from Table 12.13 is $W^t = (0.31, 0, 0.60, 0.50, 1, 0.94)^T$ and the normalized weight vector is $W(0.09, 0, 0.18, 0.15, 0.30, 0.28)^T$. The same systematic approach is considered for other evaluations and the weights obtained from Tables 12.14, 12.15 12.16, 12.17, 12.18, 12.19 are:

Table 12.4	$(0.22, 0.22, 0.22, 0, 0, 0.33, 0)^T$
Table 12.5	$(0.27, 0.46, 0, 0, 0, 0.27, 0)^T$
Table 12.6	$(0, 0, 0, 0.25, 0.25, 0.25, 0.25)^T$
Table 12.7	$(0.07, 0.4, 0.4, 0, 0, 0.07, 0.07)^T$
Table 12.8	$(0, 0.25, 0, 0.25, 0.25, 0.25, 0.25)^T$
Table 12.9	$(0.39, 0, 0, 0.18, 0.18, 0.26, 0)^T$

Finally, weights of the alternatives with respect to the goal are given in Table 12.20 and ranking of the alternatives based on overall results is given in Table 12.21. As the result of the evaluation, the three Biomass alternatives are the best ranked by

Table 12.21 Ranking of alternatives

Alternative	
A_6	0.23
A_4	0.17
A_5	0.17
A_2	0.15
A_1	0.14
A_3	0.08
A_7	0.06

the method, and the Co-combustion in conventional central plant ($P \geq 50Mw$) is selected as the best one with a priority weight of 0.23.

12.3 Concluding Remarks

When using FAHP the decision-maker is only asked to give judgments about either its preference of one alternative on one criterion against another or the relative importance of one criterion against another. However, when the number of alternatives and criteria grows, the method requires complex computations and the process becomes cumbersome. The method has been used by Wang et al. [13] for trigeneration systems selection and evaluation and by Jaber et al. [14] to evaluate space heating systems used in Jordan based on the fuzzy sets and analytical hierarchy process. Kahraman et al. [15] use two fuzzy multicriteria decision-making methodologies, FAHP and Fuzzy axiomatic design, to select the most appropriate renewable energy alternative for Turkey, Kaya and Kahraman [16] propose and integrated VIKOR-FAHP in order to determine the best renewable energy alternative for Istanbul. Talinli et al. [17] present a comparative analysis of three different energy production process (EPP) scenarios for Turkey using FAHP.

References

1. Saaty TL (1980) The analytical hierarchy process. McGraw-Hill, NY
2. Hu HA, Hsu CW, Kuo TC, Wu WC (2009) Risk evaluation of green components to hazardous substances using FMEA and FHA. Expert Syst Appl 36:7142–7147
3. Buckley JJ (1985) Fuzzy hierarchical analysis. Fuzzy Set Syst 17(3):233–247
4. Boender CGE, de Grann JG, Lootsma FA (1989) Multicriteria decision analysis with fuzzy pairwise comparison. Fuzzy Set Syst 29:133–143
5. Cheng CH (1996) Evaluating naval tactical missile systems by FAHP based on the grade value of membership function. Eur J of Oper Res 96(2):423–443
6. Van Laarhoven PJM, Pedrycz W (1983) A fuzzy extension of Saaty's priority theory. Fuzzy Set Syst 11:229–241
7. Buyukozkan G, Kahraman C, Ruan D (2004) A fuzzy multicriteria decision approach for software development strategy selection. Int J Gen Syst 33(2–3):259–280

8. Zeng J, An M, Smith NJ (2007) Application of a fuzzy based decision making methodology to construction project risk assessment. Int J Proj Manage 25:589–600

9. Chang DY (1992) Extent analysis and synthetic decision optimization techniques and applications (vol. 1). World Scientific

10. Chang DY (1996) Applications of the extent analysis method on FAHP. Eur J of Oper Res 95(3):649–655

11. Tran L, Duckstein L (2002) Comparison of fuzzy numbers using a fuzzy distance measure. Fuzzy Set Syst 130:331–341

12. Chen LS, Cheng CH (2008) Selecting IS personnel using anking fuzzy number by metric distance method. Eur J of Oper Soc 160(3):803–820

13. Wang JJ, Jing YY, Zhang CF et al (2008) A fuzzy multi-criteria decisionmaking model for trigeneration system. Energ Policy 36(10):3823–3832

14. Jaber JO, Jaber QM, Sawalha SA, Mohsen MS (2008) Evaluation of conventional and renewable energy sources for space heating in the household sector. Renew Sust Energ Rev 12(1):278–289

15. Kahraman C, Kaya I, Cebi S (2009) A comparative analysis for multiattribute selection among renewable energy alternatives using fuzzy axiomatic design and fuzzy analytic hierarchy process. Energy 34:1603–1616

16. Kaya T, Kahraman C (2010) Multicriteria renewable energy planning using an integrated fuzzy VIKOR and AHP methodology: the case of Istanbul. Energy 35(6):2515–2527

17. Talinli I, Topuz E, Mehmet UA (2010) Comparative analysis for energy production processes (EPPs): sustainable energy futures for Turkey. Energ Policy 38(8):4479–4488

Chapter 13
The Shapley Value

In fuzzy Multi-criteria decision-making problems, the ranking of alternatives must take into account their fuzzy scores in all criteria, the weights assigned to each decision criterion, the possible difficulties of comparing two alternatives when one is significantly better than the other on a subset of criteria, but much worse on at least one criterion from the complementary subset of criteria and the decision-maker's attitude toward the risk associated with evaluation. Therefore, the relationships among criteria are crucial for adequate treatment of fuzzy decision-making because they reflect the structure of interaction among the criteria and represent decision-maker's preferences of the criteria. The method presented in this chapter produces an evaluation of the marginal contribution of each of the decision criteria to the overall goal reflecting the relationships among the decision criteria and decision-makers' preferences.

13.1 The Shapley Value

The global importance of a particular criterion is not solely determined by that particular criterion but also by the value of all other criteria considered in the evaluation process [1]. Let us consider a set of decision criteria as $C = [c_1, c_2, \ldots, c_n,]$ and μ a fuzzy measure on C such that

$$\mu(C) = \sum_i \mu_i(c_i) \text{ for } i = 1, 2, \ldots, n \qquad (13.1)$$

with $\mu(\Phi) = 0$ and $\mu(C) = 1$, where $\Phi =$ null set and $\mu(C)$ is the weight or the importance value of the criterion c_i. To determine the expected marginal contribution of a particular criterion (player) to the overall goal, the importance index or Shapley value of criterion c_i with respect to μ is defined as:

$$\mu(C_i) = \sum \frac{(N - A)!(A - 1)!}{N!} [\mu(A) - \mu(A - C_i)] \qquad (13.2)$$

where N is the number of decision criteria and A is any combination of decision criteria containing criterion c_i. Thus, in addition to the usual weights on criteria

J. R. San Cristóbal Mateo, *Multi-Criteria Analysis in the Renewable Energy Industry*, Green Energy and Technology, DOI: 10.1007/978-1-4471-2346-0_13, © Springer-Verlag London Limited 2012

Table 13.1 Fuzzy numbers for linguistic variables

Definition	Fuzzy number
VG (Very Good)	(0.8 0.9 1.0 1.0)
G (Good)	(0.6 0.7 0.8 0.9)
AA (Above Average)	(0.5 0.6 0.7 0.8)
A (Average)	(0.4 0.5 0.5 0.6)
BA (Below Average)	(0.2 0.3 0.4 0.5)
P/LI (Poor/Low Important)	(0.0 0.0 0.1 0.2)

taken separately, weights on any combination of criteria must also be defined in Multi-criteria Decision-Making problems. The method proposed by Shapley [2] consists of the following steps.

Step 1. Establishing weights for subcriteria

Consider a decision problem with criteria and subcriteria. First, the decision-maker expresses, as fuzzy numbers, the importance weights assigned to each of subcriteria according to Table 13.1. Next, these values are defuzzied and then normalized.

Step 2. Calculating the Shapley value for subcriteria

In order to calculate the Shapley value for subcriteria (Eq. 13.2), decision-makers are asked to assign the weight to the combinations of subcriteria, based on the weights established for subcriteria. Thus, $w(c_{11}, c_{12})$ represents the average weight assigned to the combination of subcriteria 1 and 2 and $w(\Phi)$ is the weight to the null set of subcriteria. These subcriteria do not compensate each other, that is, a bad score on one criterion cannot be adjusted by a good score on any other complementary criteria. However, in practice, these subcriteria by and large corporate each other, that is, an increase (or decrease) in the degree to which one criterion is satisfied often increases (or decreases) the degree to which another criterion is satisfied. Therefore, the weight assigned to the combination of any two criteria must be at least equal to the sum of their individual weight assigned separately [2]. That is,

$$w(c_1, c_2) \geq w(c_1) + w(c_2) \tag{13.3}$$

Step 3. Rating of performance of alternatives

To know the scores for different alternatives against each of the different criteria, alternative's performances on each of the subcriteria are to be rated by the decision-makers.

Step 4. Calculating the Shapley value for decision criteria

Decision-makers are asked to assign the weights to each of the decision criteria and combinations of them, then the Shapley value of each decision criterion is calculated using Eq. 13.2. The values indicate the expected marginal contribution of global importance of each decision criterion to the overall goal.

Table 13.2 Criteria and subcriteria

Criteria	Subcriteria
c_1 Economic	c_{11} Investment cost
	c_{12} Net present value
	c_{13} Payback period
c_2 Technological	c_{21} Reliability
	c_{22} Efficiency
	c_{23} Maturity
c_3 Environmental	c_{31} Emissions avoided
	c_{32} Noise
	c_{33} Land use

Table 13.3 Ranking of subcriteria

Economic		Environmental		Technological	
c_{11}	VG	c_{21}	G	c_{31}	G
c_{12}	G	c_{22}	G	c_{32}	A
c_{13}	G	c_{23}	AA	c_{33}	G

Step 5. Ranking of alternatives

In order to rank the alternatives, the Total Score (TS) for an alternative can be calculated as follows [3]:

$$TS = \sum x_k \mu(c_k) \text{ for } k = 1, 2, \vdots, n \tag{13.4}$$

where $\mu(c_k)$ is the Shapley value on subcriterion k, and x_k is the score of the alternative against the subcriterion k.

13.2 Application

In this section, in order to show the use of the Shapley value, the selection of a renewable energy project for electric generation is considered. Four alternatives are eligible: Solar (A_1), Wind (A_2), Hydropower (A_3) and Biomass (A_4). The four alternatives will be evaluated taking into account three decision criteria (Economic, Technological, and Environmental) which consists of the subcriteria shown in Table 13.2. The importance weights assigned by the decision-maker to each of these subcriteria are shown in Table 13.3.

In order to calculate the crisp scores (defuzzied value of a trapezoidal fuzzy number) for subcriteria, the following equation was proposed by Kaufmann and Gupta [4]:

$$e = \frac{(x_1 + x_2 + x_3 + x_4)}{4} \tag{13.5}$$

Table 13.4 Crisp scores for subcriteria

Economic		Technological		Environmental	
Subcriterion	Crisp score	Subcriterion	Crisp score	Subcriterion	Crisp score
c_{11}	0.925	c_{21}	0.750	c_{31}	0.750
c_{12}	0.750	c_{22}	0.750	c_{32}	0.500
c_{13}	0.750	c_{23}	0.650	c_{33}	0.750

Table 13.5 Normalized weights for subcriteria

Economic		Technological		Environmental	
Subcriterion	weight	Subcriterion	weight	Subcriterion	weight
c_{11}	0.381	c_{21}	0.349	c_{31}	0.375
c_{12}	0.309	c_{22}	0.349	c_{32}	0.250
c_{13}	0.309	c_{23}	0.302	c_{33}	0.375

Table 13.6 Weights assigned to different combinations of subcriteria

Economic		Technological		Environmental	
Combination	weight	Combination	weight	Combination	weight
$w(c_{11}, c_{12})$	0.80	$w(c_{21}, c_{22})$	0.75	$w(c_{31}, c_{32})$	0.80
$w(c_{11}, c_{13})$	0.75	$w(c_{21}, c_{23})$	0.80	$w(c_{31}, c_{33})$	0.75
$w(c_{12}, c_{13})$	0.80	$w(c_{22}, c_{23})$	0.75	$w(c_{32}, c_{33})$	0.80

The crisp scores (defuzzied valued) for subcriteria are shown in Tables 13.4 and 13.5 shows the normalized weights for each subcriterion obtained by dividing scores of each c_{ik} by $\sum_{k=1}^{3} c_{ik}$.

Let us assume that the weights assigned to different combinations of subcriteria are as presented in Table 13.6.

Using Eq. 13.2 the Shapley value for subcriteria can be calculated as follows:

1. When A is c_{11} alone

$$\mu(c_{11})_1 = \frac{(3-1)!(1-1)!}{3!}[w(c_{11}) - w(\Phi)] = 0.127$$

2. When A is the combination of c_{11} and c_{12}

$$\mu(c_{11})_2 = \frac{(3-2)!(2-1)!}{3!}[w(c_{11}, c_{12}) - w(c_{1,2})] = 0.082$$

3. When A is the combination of c_{11} and c_{13}

$$\mu(c_{11})_3 = \frac{(3-2)!(2-1)!}{3!}[w(c_{11}, c_{13}) - w(c_{1,3})] = 0.073$$

4. When A is the combination of c_{11}, c_{12} and c_{13}

$$\mu(c_{11})_4 = \frac{(3-3)!(3-1)!}{3!}[w(c_{11}, c_{12}, c_{13}), -w(c_{1,2}, c_{13})] = 0.066$$

Table 13.7 Shapley value for subcriteria

$\mu(c_{11}) = 0.349$	$\mu(c_{21}) = 0.349$	$\mu(c_{31}) = 0.346$
$\mu(c_{12}) = 0.338$	$\mu(c_{22}) = 0.324$	$\mu(c_{32}) = 0.308$
$\mu(c_{13}) = 0.313$	$\mu(c_{23}) = 0.326$	$\mu(c_{33}) = 0.346$

Table 13.8 Alternative's performance for subcriteria

	Subcriterion			Subcriterion			Subcriterion		
	c_{11}	c_{12}	c_{13}	c_{21}	c_{22}	c_{23}	c_{31}	c_{32}	c_{33}
A_1	VG	G	G	G	A	A	AA	G	G
A_2	AA	VG	AA	G	AA	VG	G	AA	VG
A_3	G	G	A	AA	VG	AA	G	A	G
A_4	VG	G	AA	G	VG	G	G	VG	AA

Table 13.9 Crisp scores

	Economic			Technological			Environmental		
	c_{11}	c_{12}	c_{13}	c_{21}	c_{22}	c_{23}	c_{31}	c_{32}	c_{33}
A_1	0.925	0.750	0.750	0.750	0.500	0.500	0.650	0.750	0.750
A_2	0.650	0.925	0.650	0.750	0.650	0.925	0.750	0.650	0.925
A_3	0.750	0.750	0.500	0.650	0.925	0.650	0.750	0.500	0.750
A_4	0.925	0.750	0.650	0.750	0.925	0.750	0.750	0.925	0.650

Now, Shapley value for subcriterion c_{11} can be obtained as:

$$\mu(c_{11}) = \mu(c_{11})_1 + \mu(c_{11})_2 + \mu(c_{11})_3 + \mu(c_{11})_4 = 0.349$$

Similarly, Table 13.7 shows the Shapley value for the rest of subcriteria

Table 13.8 shows the linguistic assessment of the alternative's performance on each of the subcriteria and Table 13.9 shows the crisp scores on each of the subcriteria obtained using Eq. 13.3.

Total scores for alternatives on criterion c_1 are calculated as follows:

$$\begin{bmatrix} 0.925, 0.750, 0.750 \\ 0.650, 0.925, 0.650 \\ 0.750, 0.750, 0.500 \\ 0.925, 0.750, 0.650 \end{bmatrix} \begin{bmatrix} 0.349, 0.338, 0.313 \end{bmatrix} = \begin{bmatrix} 0.811, 0.743, 0.672, 0.780 \end{bmatrix}$$

Thus, Table 13.10 shows the scores for different alternatives against each of three different criteria.

Let us assume that the weight assigned by the decision-maker to each of the decision criteria and combinations of them for this evaluation process are the weights shown in Table 13.11. The Shapley value of each decision criterion calculated using Eq. 13.2 is shown in Table 13.12.

Scaling these values by a factor $N = 3$, an importance index greater than 1 indicates a criterion more important than the the average. From Table 13.12, it is clear

Table 13.10 Score matrix for alternatives

Alternatives	c_1	c_2	c_3
A_1	0.811	0.587	0.715
A_2	0.743	0.775	0.780
A_3	0.672	0.739	0.673
A_4	0.780	0.807	0.769

Table 13.11 Weights assigned to decision criteria and combinations of them

$w(c_1) = 0.4$	$w(c_1, c_2) = 0.70$	$w(\Phi = 0)$
$w(c_2) = 0.3$	$w(c_1, c_3) = 0.85$	$w(c_1, c_2, c_3) = 1.0$
$w(c_3) = 0.3$	$w(c_2, c_3) = 0.90$	

Table 13.12 Shapley value for decision criteria

Shapley value	0.325	0.300	0.375
Scaled shapley value	0.975	0.900	1.125

Table 13.13 Ranking of alternatives

Score	0.783	0.766	0.708	0.692
Rank	A_4	A_2	A_1	A_3

that the decision-maker attaches more weight to criterion 3 than any other criterion in the selection process. Using Eq. 13.4, the overall score for each alternative can be calculated

$$\begin{bmatrix} 0.811, 0.587, 0.715 \\ 0.743, 0.775, 0.780 \\ 0.672, 0.739, 0.673 \\ 0.715, 0.807, 0.769 \end{bmatrix} \begin{bmatrix} 0.325, 0.300, 0.375 \end{bmatrix} = \begin{bmatrix} 0.708 & 0.766 & 0.692 & 0.784 \end{bmatrix}$$

Hence, the overall performance score for alternatives and their ranking are as shown in Table 13.13. Alternative A_4, the Biomass plant is the best ranked and therefore, the most preferred alternative.

13.3 Concluding Remarks

Historically, the first application of the Shapley value was cost sharing or budgeting in situations in which multiple users contribute resources and share the costs or profits of the ensuing activities, but since then the method has been used by researchers in cost allocation problems [5], water resource allocation [6] and to evaluate the efficiency in the exploitation of hydroelectric plants [7]. In energy and environmental analysis, the method has been used by Albretch et al. [8], Ang et al. [9] and Wu et al. [10].

References

1. Shing D, Tiong RLK (2005) A fuzzy decision framework for contractor selection. J Constr Eng M ASCE 131(1):62–70
2. Shapley LS (1953) A value for n-person games contribution to the theory of games. In: Kuhn HW, Tucker AW (eds) Annals of mathematics studies. vol II. Princeton University Press, Princeton, pp 307–317
3. Hwang CL, Yoon K (1981) Multiple attribute decision making: methods and applications. Springer, New York
4. Kaufmann A, Gupta MM (1988) Fuzzy mathematical models in engineering and management science. North-Holland, Amsterdam
5. Krus L, Bronisz P (2000) Cooperative game solution concepts to a cost allocation problem. Eur J Oper Res 122(2):258–271
6. Wang L, Fang L, Hipel KW (2008) Basin-wide coperative water resources allocation. Eur J Oper Res 3(1):798–817
7. Sant'Anna LA, Sant'Anna AP (2008) A probabilistic approach to evaluate the exploitation of the gepgraphic situation of hydroelectric plants. Energ Policy 36(7):2320–2329
8. Albretch J, Francois D, Schoors K (2002) A Shapley decomposition of carbon emissions without residulas. Energ Policy 30:727–736
9. Ang BW, Liu FL, Chew EP (2003) Perfect decomposition techniques in energy and environmental analysis. Energ Policy 31:1561–1566
10. Wu JH, Chen YY, Huang YH (2007) Trade pattern change impact on industrial CO_2 emissions in Taiwan. Energ Policy 35(11):5436–5446

Index